GOLD FEVER

Incredible Tales of the Klondike Gold Rush

RICH MOLE

VICTORIA · VANCOUVER · CALGARY

Heritage House Publishing Company Ltd.
www.heritagehouse.ca

Library and Archives Canada Cataloguing in Publication
Mole, Rich, 1946–
    Gold fever: incredible tales of the Klondike gold rush / Rich Mole.—1st Heritage House ed.

ISBN 978-1-894974-69-1

    1. Klondike River Valley (Yukon)—Gold discoveries. I. Title.

FC4022.3.M65 2009   971.9'102   C2008-908131-5

Originally published 2006 by Altitude Publishing Canada Ltd.

Library of Congress Control Number: 2009920316

Series editor: Lesley Reynolds.
Cover design: Chyla Cardinal. Interior design: Frances Hunter.
Cover photo: University of Washington Libraries, Special Collections, UW7326. Interior photos: Washington State Historical Society, pages 30 and 90; University of Washington Libraries, Special Collections, pages 65 (Hegg 2159), 76 (La Roche 100) and 99 (Goetzman 620); Provincial Archives of British Columbia, page 84; Dawson City Museum and Historical Archives (#984R-216-59, Goetzman), page 103; Glenbow Archives (NA-2615-11), page 105. Map: Scott Manktelow.

Mixed Sources
Cert no. SW-COC-001271
© 1996 FSC
FSC

The interior of this book was produced using 100% post-consumer recycled paper, processed chlorine free and printed with vegetable-based inks.

Heritage House acknowledges the financial support for its publishing program from the Government of Canada through the Canada Book Fund (CBF), Canada Council for the Arts and the province of British Columbia through the British Columbia Arts Council and the Book Publishing Tax Credit.

BRITISH COLUMBIA
ARTS COUNCIL
Supported by the Province of British Columbia

The Canada Council | Le Conseil des Arts
for the Arts | du Canada

12 11 10   2 3 4 5

Printed in Canada

Contents

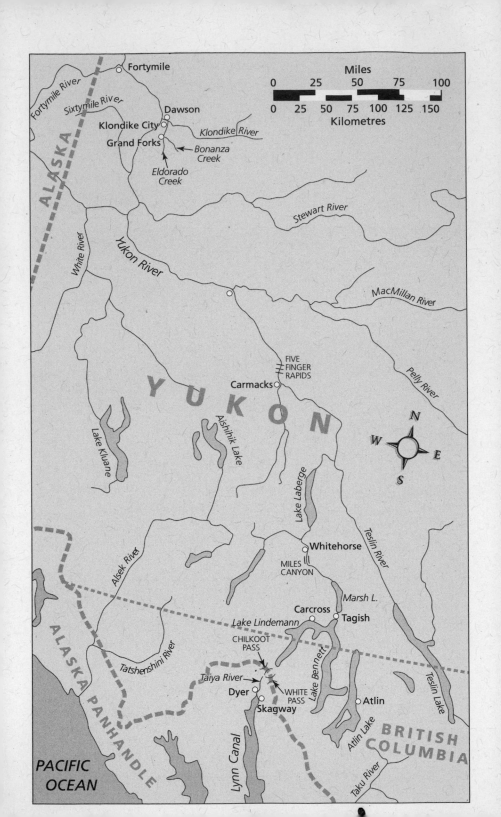

Prologue

This is the Law of the Yukon, that only the Strong shall thrive,
That surely the Weak shall perish, and only the Fit survive.
Dissolute, damned and despairful, crippled and palsied and Slain,
This is the Will of the Yukon—Lo! How she makes it plain!
— Robert Service, "The Law of the Yukon"

IT WAS THE WINTER OF 1893. *In one of Fortymile's saloons, Jim Washburn watched as the prospector sitting across from him fanned his cards out on the tabletop, looked up and grinned. The Fortymile tough guy sat in stunned silence. Washburn could have sworn he had his opponent pegged. It was inconceivable that he had read him wrongly. Where had that hand come from all of a sudden?*

Dropping his cards, Washburn reached across his belt. The glint of the knife blade flashed in the lamplight as he lunged across the heavy wooden table, spilling poker chips left and right. The man across the table stared open-mouthed at the spreading line of red staining his torn shirt. He looked up and took in Washburn's taunting leer. Then, without rising from his chair, the man calmly reached down, lifted up his revolver, pointed it at Washburn and squeezed the trigger.

The ears of every man in the saloon rang with the report. When the smoke drifted away, Washburn was writhing wordlessly on the plank floor, hands clutching his hip, blood seeping between his fingers.

The time for law and order in the Yukon had come at last.

1

Keeping the Faith

ALMOST TWO DECADES BEFORE THE great Klondike Gold Rush made the Yukon known around the world, this mountainous and inhospitable wilderness in the extreme northwest corner of Canada's vast North-West Territories was virtually unknown. The early explorers and Hudson's Bay Company traders had come and gone. But in the early 1880s, another small but steady trickle of hardened, lonely men headed north. They were not trappers or traders, but prospectors.

In the wake of earlier gold rushes in the Fraser River, Cariboo and Cassiar regions of British Columbia, and others in South Dakota, Nevada and Arizona, restless men were goaded on by the lure of easy riches. They travelled up Canada's west coast, hiked east through the mountain passes of the new American territory of Alaska and ventured into the Yukon. These taciturn individuals endured continual disappointment and desperate hardship in their obsessive quest for the "motherlode." Often having little to show for years of effort, the tenacious prospectors seemed nourished by faith alone—faith that they would one day strike it rich.

Fort Reliance, Yukon River
September 1882

The dull thud of axes was music to trader Jack McQuesten's ears. Not far from his Alaska Commercial Company trading post, cabins were taking shape. Instead of merely stopping briefly on their way to "the outside," as the prospectors referred to the civilized world to the south, twelve prospectors had decided to winter over at the post. That was good for business. The men opened accounts at the post. Jack wasn't worried about when they would pay these accounts. When a man gave his word, you trusted him. In the Yukon, that's how business was done. On trust. On faith.

Big, beefy Jack McQuesten stood watching the men carrying spruce logs; he stroked the long, blond moustache that hid his smile. Having the prospectors stay for the winter meant more than good business. It meant good company. Now there would be somebody to talk with until spring breakup. Of course, there was the missus. Katherine was an educated Koyukuk; she spoke English well, but talking to her wasn't the same as conversing with men, especially men from the outside. Take old William Moore, a veteran of the BC rushes in the Cariboo and Cassiars. Moore had made a $90,000 fortune. He didn't make his money panning nuggets from the creeks. He had made it running a fleet of steamboats! It wasn't every day you shared a bottle with a riverboat captain. Moore had lost his fortune when the gold ran out. Now he was just like the rest of them—a footloose dreamer.

For the first time in years, this winter there would be cards, conversation and laughter with white men.

Dyea Inlet, Alaska
Spring 1886

Keish, a young, powerful member of the Tagish First Nation, spotted the large brown bear not far from where the Taiya River ran into the ocean. The beast ambled leisurely out of the woods and into the warmth of the open beach. Keish stopped at once and held his breath. Had the bear seen him? Head lowered, the bear snuffled the water between the rocks. No, not yet. The wind was right. Slowly, Keish checked his rifle.

In the Tagish man's imagination, the bear was already dead. Bear meat would be very welcome around the fire. His cousin, Tagish Charlie to the whites, and his friend, George Carmack, would help him cut it up. Carmack had learned Tagish ways quickly after he and Keish had first met at Healy's trading post. Keish's sister, whom Carmack fondly called Kate, would work the skin into a nice robe. The white men were always changing Tagish names. Keish's own name meant "Lone Wolf," but he liked his white name, "Skookum Jim." Skookum meant "husky." Keish had another thought. Killing this bear would be a good story to share with his large family when they returned to their Tagish Lake village.

A throaty growl cut through Keish's reverie and his head snapped up. The bear was already racing down the beach toward the man who had dared to interrupt its feeding. There was not a second to lose. Keish raised his rifle quickly and fired. Had he hit it? If he had, the bullet had merely spurred the bear on more furiously. Keish levered another cartridge home and fired again. The bear's heavy paws made white spray of the shallows as it moved relentlessly through the water, closing the gap between himself and the puny human. There was barely time for one more shot. Keish's rifle bucked, and as the smoke cleared, the roaring bear was on him.

Stunned by the force of the bear's first blow, Keish spun away, hearing the sound of ripping fabric as his shirt sleeve shredded, feeling sudden, burning pain as the animal's claws racked his arm. If he was going to die, Keish thought, this was the best way to go. Keish rose up, extending his weapon above his head. Leaning forward with all his power, and with a roar of his own, Keish rammed the rifle's barrel as hard and as deep as he could into the bear's cavernous mouth.

Roaring in agony, the bear fell back, wrenching the rifle from Keish's grasp. The furious animal advanced again, the rifle jerking grotesquely from its mouth, blood splattering the sand. The bear swiped at him once more. Keish dodged, but not quickly enough. He felt claws open the flesh of his other arm. The mangled, bloody rifle fell to the ground and the bear lunged forward again. Keish darted away. The long claws swiped the empty air. Then the bear paused, gave a convulsive shudder and collapsed heavily on the wet sand.

Keish didn't hesitate. Falling to his knees, he wrenched a half-buried stone from the sand. Balancing his new weapon precariously above his head, he ran toward the fallen animal and brought the stone down as hard as he could on the bear's head. Chest heaving, blood running down his arms, Keish warily circled the motionless bear. It was over. Keish collapsed to his hands and knees. This would be a story to tell the others: Keish and the bear on the beach—a *skookum* story, a *Skookum Jim* story!

Over 1,300 kilometres south of the deserted beach, in the city of Victoria on Vancouver Island, events were unfolding that would have a life-changing impact on Skookum Jim.

Victoria, British Columbia
April 1887

Two Canadian government surveyors, the bearded little hunchback, George Dawson, and his young companion, William Ogilvie, had arrived in Victoria. Under orders from Minister of the Interior Thomas White, Ogilvie and Dawson were in BC's capital just long enough to hire labourers and buy supplies for their Yukon expedition. With more prospectors climbing the Chilkoot Pass, the Canadian government in Ottawa had finally realized there would be hell to pay if it didn't have a system for determining whether claims were registered with the United States or the Dominion of Canada. Canada had to protect its interests— and its citizens. It needed to establish an international boundary, and defining that boundary had brought Dawson and Ogilvie across the country. This would be William Moore's ticket north. Moore, the failed but unflagging goldfield entrepreneur, strode down the sloping street, dodging pedestrians, his mind revolving feverishly. He had waited almost three years for this moment, ever since he had returned to his wife in gentle, complacent Victoria.

"I feel convinced," Moore had written the BC lieutenant-governor, "that there are large deposits of gold at the head-waters of the Yukon River." It could be "a source of wealth to the province and of great commercial value." For a grant of $6,000, Moore would be happy to do the exploring. True, it was a significant sum of money, but a mere pittance when one considered the potential. Moore's letter had been ignored.

The previous fall, two Yukon prospectors had discovered the very first course gold in the area on the Fortymile River, about 14 kilometres west from its mouth on the mighty Yukon River. The news had only just reached the South. Now more men were anxious to head north. Gold fever had struck again. Moore had been so certain he would find gold in the North that he had left his sons to pan the Fortymile.

To Moore's delight, one of his sons had also undertaken another type of prospecting. Moore's son had met a Tagish trapper named Keish, but known by all as Skookum Jim. Moore saw Skookum Jim in his mind's eye—tough, broad-chested and taller than most of the Tagish. There was a better route across the mountains, according to Skookum Jim. He talked of an alternative route that would allow them to bypass the treacherous Chilkoot Pass. Moore was frantic with excitement. There *was* an easier way to Yukon gold!

It was not gold itself that set William Moore's excitement aflame. He loved gold more for the business opportunities it brought. When he worked in the Cassiars, there had been great opportunities for steamboat businesses and many government contracts to build trails. The same thing could happen in the North. Moore felt sure his faith in the Yukon was about to be rewarded. What mattered now to Moore was persuading young Ogilvie to hire him. No doubt plenty of adventurers were already competing for positions. Moore desperately needed one of those jobs. He was penniless, again. They had lost the house in Victoria, and he had a wife to support. Moore knew what Ogilvie would think: That old man? Although he was almost 66 years old, Moore didn't look or feel his age. Nobody ever babied him on a trail or a river! All he had to do was convince Ogilvie of that. Besides, at this moment, who in Victoria knew the Yukon better than William Moore?

CHAPTER

2

Dreamers
and Schemers

THE DISCOVERY OF SIGNIFICANT AMOUNTS of gold on the Fortymile River nudged the Canadian government into action. Long ignorant and uncaring about the Yukon, civil servants now had good reason to take a sudden interest in this far-flung corner of the North-West Territories. Some feared Canada's sovereignty might be at stake. This was not a new threat. Less than 20 years earlier, US expansionism and the Métis rebellions had threatened to cut the proposed nation of Canada in two. The federal government had used the construction of the Canadian Pacific Railroad—and the hasty dispatch of troops on its trains—to secure the western plains for the new nation.

On the other side of the Yukon's western border lay the

US territory of Alaska. No one could agree just where the border was located. Hardly anyone in either country really cared enough to find out. To disgusted American newspaper editors, Alaska had been merely "a dreary waste," and its purchase from Russia just 10 years earlier represented "a dark deed done at night."

The discovery of gold across Alaska's mountains was changing these perceptions. First came the dreamers, the hordes of prospectors with a glint of gold in their eyes. Spurred on by rumour and hope, the newcomers were staking claims in the Yukon and didn't even know it. Many considered the entire area part of Alaska. It would not be long before the schemers would come, men ready and willing to house, feed, clothe, entertain and transport the dreamers who dared to risk everything.

Healy and Wilson Trading Post, Dyea Inlet, Alaska May 1887

Canadian government surveyor William Ogilvie, prospector William Moore and others stepped onto the beach at the head of Dyea Inlet. The survey party had split up days before. The other government surveyor, George Dawson, headed off with a group to explore the Stikine River area. Ogilvie's group planned to climb the Chilkoot Pass, travel the lakes on the other side of the summit and make their way toward the Yukon River. However, their plans changed.

William Ogilvie received a surprise near the Taiya River,

just a short distance into the trip. Beneath the silent, forbidding Chilkoot Pass in the middle of the North Coast wilderness, a corner of transplanted civilization appeared. Not far from an isolated Native village stood the Healy and Wilson trading post, a tidy, two-storey building sporting real glass windows and a coat of bright, white paint.

Initially, store operators John J. Healy and George Dickson had seemed friendly and helpful. With survey instruments and six tons of supplies piled on the beach, Ogilvie was relieved when Healy introduced him to the spokesman for 120 Native packers. Then another surprise: Healy introduced Ogilvie to a white packer, George Carmack, and what appeared to be Carmack's adopted Native family. They included Skookum Jim and Jim's sister Kate, whom Carmack had taken as his wife.

Since his early stint as a US marine, George Carmack had become increasingly interested in the North. As a new marine, the 22-year-old Carmack had gone on an incredible trip to Sitka, Alaska. He was fascinated with the prospectors he'd come across. Their tales of quests for great riches brought back vivid boyhood yearnings. Also, the Tlingit Native peoples had captured his imagination. Carmack had taken the time to get to know the Tlingit; he was fascinated by their lifestyle and picked up their Chinook trade language surprisingly quickly. To William Ogilvie, the fact that Carmack had taken a Native wife was no surprise. The surveyor had seen many mixed marriages on the Canadian prairies. With the exception of John J. Healy's wife, the closest white woman was 160 kilometres away at Juneau and not likely to want to lead such a rough life. Kate was stronger and more experienced, packing along with the rest. In any case, Carmack seemed to have married well. Evidently, Kate was a chief's daughter.

Ogilvie asked Healy to provide him with information about this so-called new pass to the North. Were the rumours true? Ogilvie watched Healy cast a furtive glance at Dickson. Healy seemed a little less cooperative when the possibility of another route was mentioned, but his reaction didn't surprise Ogilvie or Moore. An alternative route would hurt business at his trading post nestled beneath the Chilkoot Pass. The new route was just seven kilometres down the inlet, at the bay the Native peoples called *Skagua*. The route was longer than the Chilkoot Pass, but lower and easier to travel. Did this fabled route really exist? Skookum Jim claimed that he'd travelled it himself, and Carmack backed up his story.

Ogilvie pondered his options. Now, their party was too small to split up, but the government surveyor would gladly spare Moore. The tiresome old man yammered on and on about the Yukon's endless possibilities, like some non-stop sideshow barker. Ogilvie told Moore that he and Skookum Jim could split off and explore this supposed route.

With a doubtful frown, Ogilvie handed Moore a surveyor's notebook and ordered him to keep good notes.

Moore and Skookum Jim planned to meet the rest of the party on the other side of the summit at Lake Bennett. Moore renewed his promise that he would continue on with the party when they met up, but only until reunited with his sons. Ogilvie could only hope that the family reunion would occur soon. Very soon.

* * *

Weeks later at Dyea Inlet, two shivering, dishevelled men burst through the door of Healy and Wilson's trading post and crouched over the threshold. John J. Healy and his wife watched quietly as the half-starved men—William Moore and his son—wolfed down their hot meals in silence. Healy paced the room, anxious to hear Moore's story. When the plates were clean, Moore cleared his throat and began telling his tale.

Ogilvie and Moore had reconnected after their separate excursions. Apparently, Ogilvie had found the Chilkoot Pass exhausting. Twenty-seven kilometres from the mouth of the Taiya River, the boulder-strewn pass rose almost vertically to a height of 1,050 metres. Most of it was a heart-hammering, back-breaking, three-kilometre climb. When Moore and his son later returned to the coast, coming down the Chilkoot Pass had been a horrible experience. Without food or shelter, buffeted by wind, rain and icy blasts of sleet, the numb and exhausted men had stumbled along. Ah, but the trip up the new route—the White Pass—had been much better, Moore said. Healy gave him a puzzled look.

Moore told Healy that Ogilvie had wasted no time in naming the new route. He decided to name it after Minister of the Interior Thomas White. White Pass was low and gentle—the preferred route, Moore stated. As co-owner of the Healy and Wilson trading post, Healy was not pleased to hear Moore's tale. He didn't like the sound of this White Pass; giving it a name lent the rival route a crucial legitimacy.

Five weeks after telling his story to Healy, William Moore and his son stood at Skagua, looking up the Taiya River. "Here is the future," Moore shouted, his words leaving a trail of vapour in the cool, moist air. "Before too many years, I expect to see a pack trail through that pass, followed by a wagon road." Moore paused and grinned. "I wouldn't be surprised to see a railroad through to the lakes." He saw more: a fine home, a town—his town—and a busy wharf. All he had to do was find the money to build it.

A few days later, John J. Healy and his partner watched Moore and his son wrestling logs from the forest, rolling them to the site of a future wharf. "It will never amount to anything," Healy shouted at Moore. "I feel sorry for you, wasting your time here!" Then, Healy and Wilson pushed their canoe into the bay. Moore looked up and paused to catch his breath. He wasn't about to listen to such naysayers—especially Healy, who had much to lose if most prospectors chose a route that bypassed his trading post! Moore knew that it took sweat, toil and time to make a dream reality. It took money, too. Moore was confident that he would find it.

* * *

With the discovery of the White Pass, the future of Healy's store at Dyea Inlet at the foot of the Chilkoot Pass was threatened. But Healy had a plan. Rather than spending time and energy fighting the rival White Pass, Healy decided to take a broader view. He realized big profits would be made supplying the Yukon's new prospectors at Fortymile. At the moment, the only supply firm in the whole territory was the Alaska Commercial Company. To seize this opportunity, Healy needed financial backing, and he knew where he could find it.

John Healy's trading background extended back to frontier days in Montana, when he had ridden across the border to establish the most infamous of all the Blackfoot whisky posts, Fort Whoop-Up. At the fort, Native peoples had traded buffalo robes and fox furs for whisky that Healy called "coffin varnish so strong you'll be able to shoot the injun through the brain or heart and he won't die until he's sobered up!" The whisky trade was enormously profitable until the North-West Mounted Police (NWMP) arrived and closed Whoop-Up down.

An old friend from Healy's wild Montana days, Portus B. Weare, was now a wealthy Chicago commodities dealer. Weare represented more than an old friendship; he represented financial business connections. Healy hoped that Weare and his associates would be eager for a chance to profit from the growing number of prospectors moving into the Yukon. In the summer of 1892, Healy travelled to Chicago, intent on persuading Weare and his associates that this mercantile opportunity was simply too good to pass up.

Healy's plan had been hatched years before, as he had watched Chilkoot Native peoples carrying supplies from the store over the Chilkoot Pass. The Native packers protected their interests closely. Because of their stamina and business acumen, Healy had developed a grudging respect for the Chilkoots, although he would never say so. Instead, he called them liars, cheats and thieves.

Healy decided to control the packing and charge miners a toll to use the trail. At that time, he had shrugged off the testiness of the Native packers. It was just business, Healy reasoned. You tried to beat the competition, not kill him. But Healy learned otherwise when a group of Sitka Tlingit arrived, eager to pack. A shouting match between the rival bands led to a bloodbath, with the dead scattered everywhere. None of that mattered now. What mattered was Healy's scheme: teams of packhorses, trading posts and a fleet of steamboats.

Fortymile, Yukon
Winter 1894–95

As bishop of Fortymile's Buxton Mission, William Bompas had watched in dismay as prospectors plied Native peoples with liquor. "There has been drunkenness of whites and Indians together with much danger of the use of firearms," he complained in a letter to Indian Affairs in Ottawa. With the Washburn saloon shooting, his worse fears were realized. Bompas wrote to Ottawa again asking for police protection, a force of at least ten officers. Months later, only two had arrived—Inspector Charles Constantine and Staff Sergeant Charles Brown. Constantine left after the season to deliver his report, while Brown stayed on alone.

At the same time, trader John J. Healy was also writing letters. He didn't even consider the Native peoples—he had other priorities. Healy's trip to Chicago to finance his scheme had been a stunning success. Healy and his wealthy Chicago investors had formed the North American Transportation and Trading Company (NAT&T), building posts in St. Michael and Circle City, Alaska. In the spring of 1893, as NAT&T general manager John J. Healy stood proudly beside the wheelhouse of the company's new steamboat, the *Portus B. Weare*. With its whistle blowing and twin smokestacks belching, the steamboat pulled up at Cudahy. For Healy, events were unfolding as they should. Most events, at least.

Prospectors appreciated Healy's lower prices and better selection, but not his strict payment policy. They wondered why he couldn't be more flexible, like McQuesten, who worked for the Alaska Commercial Company. However, Healy had bigger problems. It was hard to keep law and order, and he suspected the *Portus B. Weare*'s skipper of stealing. When he finally confronted the skipper, shouting led to fisticuffs. After bystanders separated the combatants, Healy fired the skipper. To settle the dispute, a miner's meeting—the crude method of dispensing justice—was convened. The prospectors quickly awarded the thieving skipper all of the fees stated in his three-year contract with Healy. Then it was drinks all around.

"The bad men of the country have settled at Fortymile Creek," Healy wrote to the minister of the interior. "These men are not miners but a class who make their living at the expense of others." The time had come for formal, impartial law and order. "We beg that you send a sufficient force of police to the Yukon so that life and property may be protected."

Healy had no illusions about slow-moving Ottawa bureaucrats. What the Yukon needed was a man of action. He addressed his next letter to the commanding officer at Fort Macleod, the man who 20 years earlier had helped shut down Healy's own prairie trading enterprise, Fort Whoop-Up. Sam Steele was the kind of man the Yukon needed, but Steele would be a long time coming.

CHAPTER

3

The Days Everything Changed

BY 1896, THE SHORT, FRANTIC gold stampede to Fortymile River was already a memory. However, each day on the banks of a dozen creeks, prospectors were still washing out just enough gold to prompt newspaper stories in southern cities. The reports inspired a brave—or desperate—few to leave starvation wages or unemployment behind to try their luck in the North. Upriver from the mouth of the Fortymile, and in a number of nearby creeks, prospectors were pouring barely enough gold into caribou-hide pokes to keep them panning and digging. They trusted that someone soon would make the next big find. Each one harboured the secret hope that he would be the one.

Traders Jack McQuesten, Joe Ladue and John Healy

were hoping the same thing. McQuesten and Ladue continued to advance the prospectors' grubstakes against the day someone would strike it rich and rescue them all.

Ogilvie, Sixtymile River, Yukon
Fall 1894

The weary, dejected prospectors told the man at Sixtymile River they were from Colorado. Joe Ladue, part owner of the trading post and sawmill, looked closely at the trio. Like them, Joe Ladue had been a rootless wanderer, leaving his sweetheart in New York to seek gold throughout North America. For him, finding a fortune meant instant acceptance from his girl's wealthy family. Still, Ladue had finally had it with the creeks. There must be another way to make it out here, he thought. Ladue had a way about him: a real gift of the gab. Prospectors fed on enthusiasm like that. Earlier, trader Jack McQuesten had recognized this gift in Ladue and helped set him up at Sixtymile. The post had been named Ogilvie, after the Canadian government surveyor. Leading the weary prospectors into the store, Ladue gave them one of his usual rousing speeches, but only one man, Robert Henderson, seemed to respond with interest. He asked too many questions for someone ready to go back home.

There were great prospects on Indian River, Ladue told Henderson. Indian River was less than 30 kilometres down the Yukon. The more Ladue smiled and talked, the more intrigued Henderson became. Then, Ladue asked a few questions of his own. Henderson had left his family in Colorado, but that was only his most recent home. A lightkeeper's son from Nova Scotia, Henderson had sailed the world in an endless quest for gold, spending the last 14 years in Colorado alone. It was obvious to Ladue that Henderson had the tenacity every prospector needed but few possessed.

"Whaddya need?" Ladue asked. Henderson pulled a dime from his coat pocket and looked back at him. Robert Henderson was willing to take one more gamble and hoped Ladue was, too.

"Let me prospect for you," he told Ladue. "If it's good for me, it's good for you. I'm a determined man. I won't starve," he promised. If a dedicated prospector needed a grubstake, just like his Alaska Commercial Company colleague, Ladue was happy to oblige.

Seattle, Washington
1895–96

Thomas Lippy, the Seattle YMCA secretary, reflected on the irony of his situation. Thanks to a knee injury, his new office job paid more than his former position as a phys. ed. instructor. With so many out of work, Lippy knew he should count his blessings. Yet he yearned to do something more than push a pen. Just what that was, he wasn't sure.

Stories about Yukon prospectors had caught Lippy's eye. Men were finding gold up there. He couldn't get the stories out of his mind. Worse, he found he wanted to be a part of those stories. Lippy could feel the pull of the North, and in some indefinable way, he knew this was his chance. The Yukon was physically punishing, Lippy had heard. That didn't concern him, but he knew that the longer he spent sitting behind a desk, the faster his muscle tone would disappear. He couldn't afford to wait much longer to make a decision.

Lippy had checked out the cost of Alaska outfits. Money was a problem. He and his wife, Salome, would have to borrow for what they needed. Still, they were young, without children. Lippy marshalled all his arguments and prepared to broach the subject of going north with Salome. When he did, he would avoid the word "hunch." Using a word like that was no way to win over one's wife.

Beneath King Solomon's Dome, Yukon
1896

For two years, Robert Henderson had endured black clouds of summertime insects. He had suffered freezing, snow-driven winter winds. He had worn out boots and pants, and he had eaten through all of his supplies. He had suffered temporary snow blindness and another time barely managed to make it ashore when his skin boat capsized. Once, he slipped on a tree he had just felled, impaling his leg on a jagged branch. Henderson had hobbled off to lie alone in his tent for more than 20 days of agony, not sure he would survive. He had used strips of bacon as a poultice, throwing them to the wolves when they became too rancid. The leg still bothered him, but at least he was alive, and he always found enough gold to pay Joe Ladue for the next outfit— sometimes alone, sometimes with a partner.

In the early summer of 1896, Henderson had broadened his search for gold in the Yukon, wandering down to a little creek on the northern side of the mountain called King Solomon's Dome. He dipped his pan into the waters, washing away the gravel to reveal eight cents worth of gold. That was good, *very* good. He named the creek Gold Bottom and then hurried back to persuade three other luckless prospectors to return with him to the creek. Within weeks, the four men had shaken $700 worth of gold into their pokes.

Fortymile, Yukon
1896

Around the same time as Henderson was prospecting at Gold Bottom Creek, George Carmack had an incredibly vivid dream that made further sleep impossible. In his dream, he reached down in blue-green water to grasp a salmon. This was no ordinary fish. The salmon's scales were made of glistening gold flakes and gold pieces covered its eyes. The only blue-green waters George knew about flowed into the Yukon 70 kilometres south from a salmon stream that prospectors called the Klondike. Before morning, Carmack told his wife that they would fish for salmon on the Klondike and then prospect for gold.

On the swampy, mosquito-infested flats where the

Klondike and Yukon rivers met, Carmack constructed a willow fish weir and set out the nets. The first few days passed pleasantly enough beneath the barren mountainside called Moosehide. However, catches were few and far between. Carmack doubted they would have enough fish for winter, and the thought left him both frustrated and fearful.

An unexpected family reunion dispelled the gloom when Skookum Jim, Tagish Charlie and Charlie's younger brother bobbed down the Yukon. It had been many years since Kate had seen her brother. Skookum Jim and Tagish Charlie told them they'd tried prospecting, but nothing had come of their efforts. Even the caribou had eluded them. "Rejoin your old prospecting friend, if you want to rid yourself of evil spirits," a Tagish shaman had advised, and so they did! Carmack listened to them with interest and then shared his own vision with the newcomers.

"I'm anxious to start up the Klondike. That's why I came here," he told them. "But first I want to lay in a good supply of salmon. Then we'll go prospecting."

What if the salmon don't come? Skookum Jim asked. Carmack had an answer: they would cut trees and float logs down to the sawmill at Fortymile. "We sell the logs, buy grub, go prospecting," he said casually. They were still fishing a few days later when another visitor came calling.

Robert Henderson was returning from yet another trip for supplies at Ladue's post, where he had shared news of his findings on Gold Bottom. Henderson decided to paddle down the

George Carmack: His discovery started it all.

Yukon to the Klondike. As he poled his loaded craft past the flats at the mouth of the river, he could see people at work.

The pungent smell of smoking salmon wafted across the water. Henderson wrinkled his nose. Here was a surprise: one of the fishermen was white. "There's a poor devil who hasn't struck it," Henderson thought to himself as he slid his boat up on the mud. Why else would he stoop to fishing with the Natives? Maybe he would like to join them on Gold Bottom. As he walked closer, Henderson realized the white man in the group was George Carmack. That explained everything, thought Henderson. Carmack was known to spend most of his time with the Tagish.

"Hello," Carmack yelled, walking over to Henderson. Carmack was anxious to find out what he was up to. "I heard at Fortymile you were working for Billy Readford up on Quartz Creek. You still there?"

"Naw. Went over the divide to a little creek on the other side of the Dome. I've got a good prospect going there— named the creek Gold Bottom." Carmack became more interested. Henderson explained that he had been to Ladue's for supplies and was heading back to the creek.

"Any chance for us to stake up there?" Carmack asked eagerly.

Henderson shot a glance over George's shoulder at Skookum Jim and Tagish Charlie. "There is for *you*, George. But I don't want any damn Siwashes staking on that creek," he sneered.

As Henderson pushed his boat out into the water and paddled past the camp, Skookum Jim walked up to Carmack, his eyes narrowed—he had overheard the insult. Carmack turned to him and whispered, "Never mind. This is a big country. We'll go find a creek of our own." Telling his wife that they would be back in a week or so, the men began to pole up the Klondike.

After the first few trees had been felled and limbed, it was obvious to Skookum Jim and Tagish Charlie that it was gold—not logs—that Carmack was after. When a south fork of the Klondike called Rabbit Creek held out little promise other than as a source of timber, Carmack led his two companions through the swampy country and over the domed mountain to join Henderson's group at Gold Bottom.

Before long, supplies dwindled. Skookum Jim and Tagish Charlie asked Henderson if they could buy some of his tobacco. Henderson refused. The previous winter, he had almost starved when one of his food caches had been pilfered. As the only white man within miles, he knew who was responsible. It was clear that relations between Henderson and the Native peoples were as frosty as they had been weeks earlier. Carmack was harmless enough, Henderson felt, as long as you didn't take him too seriously. He had a reputation as a teller of tall tales.

Carmack decided to return to Rabbit Creek. True to the prospectors' code, he and Henderson promised to keep each other posted on developments.

Rabbit Creek, Yukon
August 16, 1896

No one really knows who made the find on Rabbit Creek or the exact nature of that find. Was it, as Carmack insisted, a thumb-sized gold nugget he saw wedged between slabs of bedrock? Was it the nugget Skookum Jim spied beneath the clear water while washing out a dishpan? Was it, instead, course gold that glinted in the bottom of a prospector's pan?

Laughing and yelling, Skookum Jim, George Carmack and Tagish Charlie danced wildly around the gold pan sitting on the gravel on the bank of the small, meandering stream. Inside that metal pan was gold—more than the three men had ever seen in a single prospector's pan. The pan was heavy with it. They had heard stories about gold finds like this, dreamed of it, wished for it, and now . . . there it was.

The three men calmed themselves enough to attempt some serious panning. Within minutes they were dumbstruck. They could not believe what the creek bottom revealed. Ten- or twenty-cent prospects were enough to make a prospector's heart pound. On Rabbit Creek, the average pan was yielding $4 worth of gold, an amount that a man might work for days to shake into a caribou-hide poke.

Others would need proof of the find. Carmack pried open a shotgun shell and dumped its contents onto the rocks. Then, he scooped up some of the gold and poured it carefully into the empty shell, crimping the folded cardboard tight at the top.

The find changed the men and their families forever. For Carmack, gone was any notion of living on the land. Nor would he would ever work for a trading post, fish a river or sell logs. Kate would never have to sew moccasins for white men again. The men sang all night around the fire. That night, chanting Tagish songs, it seemed to Jim that he had found the golden key that unlocked the door to life as a white man.

The next morning, when the previous night's euphoria had faded away, the question remained: who was going to stake the initial, or discovery claim? They reached a compromise. Carmack would stake and record the claim but assign Skookum Jim half interest. That done, the men discussed other claims. Carmack staked No. 1 below the discovery, from the base of hills on either side of the creek, as regulations stipulated. No. 2 below the discovery was staked for Tagish Charlie, and No. 1 above the discovery was Skookum Jim's. Carmack notched a section out of a nearby spruce tree and printed a claim notice on the tree's exposed white surface:

To Whom it May Concern:

I do, this day, locate and claim, by right of discovery, five hundred feet running up stream from this notice. Located this 17th day of August, 1896.

G.W. Carmack

CHAPTER

4

Bonanza

GEORGE CARMACK HAD BEEN IN gold country long enough to know what he had to do immediately after the claims were staked. Carving a notice on a tree was not enough. He had to formally register their claims. All three wanted to go, but Carmack persuaded Skookum Jim to stay at the creek to ward off anyone who might happen by. Soon, Carmack and Tagish Charlie were poling down the Yukon River to Fortymile. First thing in the morning, they would visit the registration office.

Not long after the pair reached the little mining settlement, news of their discovery on Rabbit Creek changed the lives of every man and woman who lived in the Yukon territory—almost overnight.

Bill McPhee's Saloon, Fortymile, Yukon
August 1896

George Carmack gulped down his first drink and motioned to the bartender for another. Clarence Berry picked up the bottle and poured the whisky carefully into Carmack's shot glass. When saloon owner Bill McPhee had agreed to take Berry on as bartender, Berry had told himself that the job would be merely temporary. Berry and his new wife, Ethel, had arrived from San Francisco only a few short months ago.

Berry thought about how he and Ethel had ended up here together. After spending two unsuccessful years in the Yukon with his brothers, Berry had returned to his fiancé to tell her some bad news. She would have to wait a while longer before they were married. But Ethel was becoming impatient and frustrated. It was bad enough that she had to convince her parents that Berry was still the man for her, gold or no gold. Now, she had to convince the man himself that it was now or never! In Clarence Berry's mind, postponing the wedding until he made his strike was the right thing to do. But not in Ethel's mind. The right thing to do was to get married right away.

Ethel's parents wondered why Berry couldn't find work closer to home. However, conditions were dire. Economic hardship continued to beset thousands of people all over the country. Everyone, it seemed, was dissatisfied and restless.

No one could predict what might happen to Berry if he went up north again. Ethel knew where she wanted to be. She wasn't afraid of the wilderness. Robust and buxom, Ethel knew she had the right physique for roughing it. So when Berry left for the Yukon in the spring, she was right there by his side. Their honeymoon was one to remember.

After meagre showings during an initial five-week prospecting trip when the couple first arrived, Berry and his brother, Fred, had given up for the season. In Berry's estimation, the Fortymile creeks, already worked steadily for nine years, were pretty well played out. So were Berry's funds. After the next spring breakup, he and Ethel would leave the motley collection of tents and log buildings that dotted the spit of land at the junction of the Yukon and the Fortymile Rivers and start prospecting elsewhere. Until then, the steady work at the saloon would enable them to purchase the outfits they would need.

Compared to prospecting, saloon work was easy and warm. In addition, Bill McPhee's was a good place to be if you wanted to stay on top of events. Clarence Berry listened closely when liquor or dance-hall girls loosened the tongues of prospectors on their sprees. This way, Berry would hear the news firsthand. For instance, he was listening when George Carmack slammed down his empty glass and turned to face the crowd in the large room.

"Boys, I've got some good news to tell you," Carmack shouted over the babble of voices and sharp clack of billiard balls. Faces turned his way and the room grew quiet. "There's a big strike up the river."

"Strike, hell," came an impatient retort. "That ain't no news. That's just a scheme of Ladue's to start a stampede up the river."

"That's where you're off, you big rabbit-eating male-mute!" Carmack yelled. "Ladue knows nothing about *this*." With a flourish, he held aloft the shotgun shell. Grinning, he stepped over to the gold scales and upended the container. A golden stream poured out. "How does that look to you, eh?" he challenged.

Boots thumped and chairs scraped as McPhee and the others gathered around the scales. Some sneered. The prevailing wisdom was that only fools believed George Carmack or that fast-talking Joe Ladue. Still, to those who figured they could tell gold's place of origin simply by its colour and texture, the pile of dust and nuggets was intriguing. This wasn't gold from Glacier, Miller or other Fortymile creeks. It didn't look like gold from the Stewart River or Indian River, either. It was gold from Rabbit Creek, Carmack announced. More skepticism followed. Joe Ladue had prospected there years before and come back empty-handed.

Derisive laughter turned to excited speculation. Clarence Berry caught Bill McPhee's eye. Berry talked low and quickly into the saloon owner's ear. McPhee sighed but nodded. If Berry needed a grubstake, McPhee would extend it. Berry might be just another fool, but he was an honest one. If things went bad, McPhee was confident that Berry would work out his debt behind the bar.

* * *

The claims registration office was not actually situated in the settlement of Fortymile itself, but inside the nearby NWMP post, Fort Constantine. It was named after the same Charles Constantine who had entered the Yukon to reconnoiter the area and determine what force of policemen would be necessary to safeguard lives and property. The summer before, Constantine and a small contingent of policemen had been hard at work constructing the most northerly military establishment in the British Empire. The site of the post was no accident. Its presence just across the mouth of the Fortymile River was a reminder to all the prospectors who lived and worked in the area's most important gold creeks that Canadian law and order had arrived in the Yukon.

* * *

George Carmack and his two Native companions poled across the mouth of the Fortymile in eerie silence. Beaching their raft, they paused in front of the gates of the post and looked down the riverbank toward Healy's store. Cudahy was deserted, much like the settlement across the river. In the darkness of the previous night, every prospector who could find a boat had poled up the Yukon to the Klondike River. Overnight, Fortymile had become a ghost town.

Inside the NWMP post, Inspector Charles Constantine listened intently to the claims recorder's story of Carmack's visit, angled his cloth cap above his right eyebrow and walked out of the office. Ramrod straight, he strode across the post's unnaturally quiet quadrangle. He had to see government surveyor William Ogilvie immediately to discuss this new development. In a day or two, when all of the prospectors came back downriver to register their claims, there would be no time to discuss anything. Nevertheless, he wanted a few minutes alone to collect his thoughts. A shotgun shell full of gold didn't mean much, but Carmack's story might. Was it true? Everyone else obviously believed it. The place was deserted. That fact made truth irrelevant.

William Ogilvie was not a happy man. Constantine couldn't blame him. Ogilvie's years of painstaking work to pinpoint the exact location of the border had finally culminated in the confirmation of the position of the international boundary on the Yukon River. The rich creeks around Fortymile appeared to be in Canadian territory. Now, Ogilvie desperately wanted to go home to his family in eastern Canada. Spring floods ended that notion. Next, Ogilvie received a fateful letter. A joint commission had been formed to finalize the creation of an international boundary, and Ogilvie had been duly appointed its Canadian representative. He was to remain in Fortymile to "await further instructions." Predictably, instructions had not yet arrived. The summer was all but over. In a few weeks, it would be too late for instructions to arrive or for Ogilvie to leave—the river would be frozen solid.

And now, how would Constantine keep the peace? After

the Mountie had met with Bishop Bompas and others dur-ing his brief visit the previous year, he had recommended a force of 40 policemen be sent north. Ogilvie had disagreed. The Fortymile rush had long since passed, he had argued. Ten men would do. Ottawa compromised and approved 20. If Carmack's find sparked another rush, as it already appeared to have done, their little contingent would be swamped. For the men who had travelled so far and toiled so hard for less than a dollar a day, quick riches would prove irresistible, and ranks would thin overnight.

Outside the wooden gates, Constantine leaned up against the palisade and stared out at the broad, flat river. The men had worked so hard to build the police post. What if it was located in a place where there was no one left to protect?

* * *

After Carmack's claim, Clarence Berry and his brother headed to Rabbit Creek. By the time they arrived, many others had already pounded claim stakes into the rocky sand of the creek banks. The Berrys wasted no time in doing the same. The brothers had to walk a good distance past Carmack's own claim to do it. Below the discovery, every inch of creek bed was already staked out. The first unclaimed spot they found turned out to be "40 Above"—40 claims above the discovery!

When he stood back and looked at his stakes straddling both sides of the creek to the base of the hills, Berry marvelled. Just hours before, this creek had been silent and empty. Now, swarms of men were at work. Before long, claim jumpers would be tossing stakes aside. There was no time to lose. Leaving his brother at the site, Berry hurried down to the Klondike.

A few hours later, after he registered his claim at Fort Constantine, Berry crossed the mouth of the river. He met briefly with Ethel, who was desperate for news. He had registered a claim, Berry told her. Now he had to load more supplies and head back up the creek to start work immediately. Months of toil lay ahead. It was impossible for him to carry everything they needed for the winter. He explained that he would be back soon, and they would discuss what to do next. Ethel watched nervously as her husband pushed off, the overloaded rowboat wallowing in the river.

After their brief stay in Fortymile, George Carmack and Tagish Charlie made their way up the Yukon River, stopping at their fishing camp at the mouth of the Klondike. There was no need for anyone to stay there any longer. They were through with fishing. They gathered up the group and launched their raft into the Klondike River. After thrashing through the underbrush from the Klondike River, Carmack, Tagish Charlie and the rest of their group met Skookum Jim on Rabbit Creek. Skookum Jim proudly showed them 85 grams of gold he had panned from his own claim, which he assumed was now duly registered in his name.

No, the claims were not registered, Carmack muttered.

The fees were $15 each and that shell only held $12. No matter: they had up to 60 days to register, he'd been told. What they needed to do now, Carmack continued, was get enough gold out of the ground to pay the registration fees and obtain their winter supplies. The men knew they had no time to lose. They quickly borrowed a whipsaw and began cutting the logs Skookum Jim had felled. The new boards were hammered into sluice boxes, which would be used to separate the gold from mounds of pay dirt.

All along the creek, the future was unfolding hourly. Before the end of the day, more than 20 claim holders met on a nearby hillside to appoint fellow prospector Dave "Two Fingers" McKay claims recorder, using a measured length of rope to define boundaries. The name "Rabbit Creek" was now a joke. The group agreed the creek should be given a more fitting name: "Bonanza."

Upper Yukon River
August 1886
On the riverbank near his trading post in Ogilvie, Joe Ladue stood in stunned disbelief. After all these years of firing up prospectors with wild tales of discoveries in the area, Ladue couldn't believe that others were the ones telling *him* tall tales. Men were streaming out of the cabins that dotted the island, pushing off in anything that floated. Was this the big one? In his heart, Ladue knew it must be so. The rafts and boats racing toward the Klondike River were all

the evidence he needed. Rabbit Creek! Who would have guessed? He dashed for his raft.

Ladue's destination wasn't Rabbit Creek. Instead, he guided his platform to the swampy ground at the mouth of the Klondike River. The deserted Carmack camp was now a rest stop for prospectors on their way up the Klondike River. The sight reinforced Ladue's conviction that this was where his own personal "strike" would be made.

"Hey, Joe," came a breathless shout. "I'm gonna need a house. Got any lumber?"

Suddenly, it wasn't the stunted trees or the haphazard smokehouse of the fishing camp that Ladue envisioned. Instead, it was a street lined with wooden buildings. He sprinted for the beach. Next, he collared a young prospector and gave him a job on the spot. While Ladue poled his raft back upriver toward his trading post, the young prospector paddled downriver toward Fort Constantine—Joe Ladue had sent him to register something he called a "townsite."

Back at Ogilvie, Ladue yelled for the staff to load every piece of lumber onto the raft. The boards weren't doing any good stacked up around the post. They had to get them to the swamp. In the meantime, Ladue ordered people to dismantle the sawmill and float the pieces down the river too. The plan fired the men into frenzied activity. At last, veteran prospector Joe Ladue was going to strike it rich! This time, he wasn't going to have to pan, dig or sluice gravel to do it.

Bonanza Creek, Yukon
September 1896

Bonanza Creek was still crowded, but now it seemed strangely quiet. Having staked their claims, many men wandered about, morose and depressed. The euphoria had vanished. Some began to drift back to Fortymile, convinced that nothing would come of the creek. Others journeyed farther down the Yukon River to Circle City, Alaska.

"For two bits I'd cut my name off the stakes," snarled one prospector. Nobody listening nearby had that much cash on them, so he resigned himself to working his worthless dirt. This was the kind of talk Alex McDonald had been hoping to hear. The deceptively slow-moving, swarthy giant from Nova Scotia studied the situation closely. Having been a land buyer for the Alaska Commercial Company, McDonald was quite happy to loiter in the area. He was waiting for the right moment to present an offer on the right property, as he had arrived too late to stake a claim himself.

* * *

For weeks, Robert Henderson had toiled away on Gold Bottom Creek, oblivious to the mounting excitement a few miles away. One day, visitors arrived from Bonanza Creek. Henderson had never heard of a creek called Bonanza. It was the new name for Rabbit Creek, the visitors explained.

"What have you got there?" Henderson asked.

"We have the biggest thing in the world," one replied.

"Who found it?" Henderson asked, in spite of himself.

"Carmack."

Henderson threw down his shovel in disgust. Carmack had beaten him! Within a few short weeks, Carmack had stumbled upon the wealth that he had spent decades searching for! When Henderson finally managed to pull himself together, he realized that he had to move fast. When Bonanza Creek was staked out, the nugget-hungry hordes would begin moving down to Gold Bottom Creek—his creek!

Fort Constantine, Yukon River
September 11, 1896

The old visionary from Victoria, Captain William Moore, finished his mail run. He dropped the sacks of mail on the floor of the office and sat down heavily, his mind a whirl. So the big strike had been made—just as he always knew it would be! He had heard the news at Dyea Inlet. As he made his way downriver on the last of his summer mail runs, Moore witnessed the frantic activity at the mouth of the Klondike River. Here, Joe Ladue was building a saloon and sawmill at his new townsite.

If Ladue was building a townsite on the Klondike River, Moore would do the same at Skagway. It meant years of seemingly endless jobs—fishing, mail delivery and piloting—were over. He couldn't wait to break the news to members of the British syndicate. They had hired him to develop the wharf and land in Skagway. They would be ecstatic at the news! The quickest way to Victoria was to board a steamer at St. Michael, Alaska.

One of the sacks of mail that Moore carried contained a letter from Ottawa addressed to surveyor William Ogilvie. Ogilvie read the letter with a sinking heart. The letter reported that the international joint commission formed to establish the final position of the border between Alaska and the Yukon had failed and been disbanded. Ogilvie's services as Canada's representative were no longer required. All these months of waiting for instructions had been for nothing! However, the letter brought some welcome news. Ogilvie had been ordered back east. He was leaving the Yukon at last. He prayed the steamer that would take him to St. Michael, Alaska, would arrive at Healy's trading post before freeze-up.

At the same time, inside his snug log home just a short walk from the post, Inspector Charles Constantine reread his letter to the police commissioner: "There is one thing for certain, unless the government is prepared to put a strong force in here next year, they had better take out what few are now here. The rush in here will exceed any previous year."

The news of the most enormous gold finds since the Cariboo Gold Rush—perhaps even since the California Gold Rush—would inevitably surface in the South. Constantine was haunted by the fear that lawlessness would overrun the Yukon. Again, he asked himself the question: how could his small band of policeman hope to maintain law and order?

5

Eldorado

ON THE SLOPE ABOVE BUSY Bonanza Creek, Austrian sheep-herder Anton Stander stood with his four companions watching the activity. Stander's group had been luckier than others. They had managed to stake a claim. Yet, the best ground had already been taken.

Gazing down the creek, Stander noticed what most others already knew but had chosen to ignore. Bonanza Creek actually forked, and the straight southern arm meandered peacefully through a ravine. The five penniless men understood something else: if something good didn't happen soon, their future was too bleak to contemplate. Nothing much was going to happen for them on the busy Bonanza Creek, they decided, so they packed up and wandered down past the fork. Their decision was to have enormous repercussions.

South Fork of Bonanza Creek
Late September 1896

William Johns cursed his luck. Inside Bill McPhee's saloon, the former newspaperman had dismissed Carmack's tale of the big strike. Johns was sadder but wiser now. He and others were camping near Carmack's claim, scouting around the area. Anton Stander came by to warm his hands and share some coffee. Johns asked Anton where he was working. He said only that his camp was "farther up." He seemed evasive. The group assumed that Stander meant "farther up on Bonanza" and left it at that.

The next morning, as the group of prospectors panned a creek, they saw one of Stander's companions. He told the group that they weren't having much luck. The men resumed their work until one of them noticed something strange about the colour of the river. "Someone's working farther up," he said. "The water's muddy." Around a bend in the creek, they came upon Anton Stander and his companions. They were all staring at Stander's pan. When they looked up at the prospectors, Johns remembered later, "They acted like a cat caught in a cream pitcher." It was too late to continue a silly pretense of disappointing returns, so Stander showed them the gold lying in his pan. The group was suitably impressed. Stander grinned and told them their very first pan had yielded more than $6 in gold. No wonder all five in the party had staked claims! Quickly, Johns and the other interlopers walked down the fork and did the same. Someone jokingly christened the creek "Eldorado." Early the next morning, as he walked to the claims recorder, another prospector told everyone along the way the happy news of strikes on the new creek.

The prospectors were careful to stake some distance from Stander's group to avoid any confrontation. However, others didn't care. For instance, claim No. 7 had already been staked illegally when two men came upon it. They simply yanked out the stakes and substituted their own.

Fortymile, Yukon
Late September 1896
Anton Stander stomped out of the Alaska Commercial Company store and down the muddy trail to Bill McPhee's saloon. He needed a drink—badly. He was in town to buy winter supplies for his group, and at the moment it looked like he would have to return empty-handed. Standing at the bar, he complained to the bartender about the new Alaska Commercial Company operator. Things had certainly changed since Jack McQuesten had moved downriver to Alaska's Circle City (he had moved to avoid paying NWMP duties). McQuesten's replacement did not believe in offering his customers unlimited credit, at least not without a guarantor.

The bartender, Clarence Berry, listened to Stander with

more than the usual sympathy. The need for money had put Berry behind Bill McPhee's bar again. Work on Bonanza Creek had been as expected: long, hard hours crouching in the holes, coughing and choking as the smoky fires thawed the permafrost, followed by back-breaking digging and hauling.

The Berrys were getting a fair return, but soon, as the temperatures plummeted, there would be no running water for the sluice boxes. Gold-washing wouldn't begin again until spring. Expenses wouldn't stop, though. Clarence Berry knew that they would eat a mountain of supplies between October and March. Happily, Berry had something Stander didn't—a paycheque.

However, in the weeks since they had staked a claim on Bonanza Creek, it became obvious that the Berrys lacked something that Stander did have: a claim on Eldorado Creek. As Berry watched the frustrated prospector nurse his drink, he began to formulate a way to help them both. Within minutes, the men were shaking hands on the deal. Berry agreed to act as Stander's guarantor. In return, Berry would trade half of his Bonanza claim for half of Stander's Eldorado claim. The two men were partners. Berry hunched that Eldorado would pay off in a big way.

* * *

Twelve sleepless hours after she had last waved goodbye to her husband, Ethel Berry had amassed five tons of supplies.

Standing on the wharf near Fortymile, she heard the whistle of the Alaska Commercial steamer *Arctic*. After supervising the loading, Ethel stepped carefully up the sloping plank to the deck of the crowded little vessel. As the boat churned upriver to the new townsite, Ethel stared up in surprise at a rare sight on the deck above. There was a young white woman standing next to a tall, muscular man. Ducking inside, wincing at the engine's deafening racket, Ethel made her way up the narrow staircase. She stepped out onto the deck and introduced herself to the couple.

"How do you do," the short, lithe woman exclaimed, as delighted as Ethel was to meet another lady this far from civilization. "I'm Salome Lippy, and this is my husband, Thomas."

Eldorado Creek, Yukon
October–December 1896

Land-buyer Alex McDonald sensed the time was right to present an offer on Eldorado Creek. The price of this particular claim certainly was right! The Nova Scotian, known in Antigonish as "Big Moose," handed the disillusioned landowner a sack of flour and a side of bacon—hardly a big risk—and site No. 30 was his. Shortly after the purchase was concluded, work on the claim uncovered a stunning pay streak that was 12 metres wide.

Others stared open-mouthed. The gold in this particular stretch of ground was clearly worth thousands. As usual, nobody could guess how rich the seam actually was. McDonald moved quickly to buy as many claims as he could. Before long, a hired hand was panning $5,000 in gold a day from the claim.

With McDonald's newfound riches came an all-consuming objective: the acquisition of as much wealth as possible. McDonald didn't have a lot of money. No problem. He simply mortgaged his existing claims and delayed payment on new ones.

In a desperate race against the elements, the frenzied activity on Eldorado Creek escalated. The enticing pull of "outside civilization" grew as hours of daylight diminished. The Yukon winter was a fierce, long, lonely ordeal that sent men mad with cabin fever. Poverty-stricken prospectors had no choice but to stay. However, many Bonanza and Eldorado claimholders were no longer poor. George Carmack had already dug out $1,400 by the end of September!

How much gold was enough to make a man leave before the snow flew—$10,000 or $20,000 worth? At very productive No. 16, two Vancouver Island coal miners decided to call it quits at the five-metre level. It was too soon, their third partner decided. He bought their shares for $50,000 each and readied himself to endure the winter.

The Berrys had no intention of leaving. Clarence Berry started burning a shaft down through the clay on the Eldorado claim he now shared with Anton Stander. In early November, Berry reached bedrock. A single pan of pay dirt revealed $57 in gold. None of the Berrys would ever go hungry again. One afternoon, Ethel came up to the claim to call her husband for supper. While she waited for him to emerge from the shaft, she idly picked out $50 in gold from the dirt at her feet.

Quickly, Berry and Stander began hiring men to haul dirt up to the surface and pile it for the spring "clean up" that would surely yield enormous riches. Paying their growing number of employees was easy. Each day, they handed them a few of the nuggets they dug out of the ground. However, keeping employees was a challenge.

"Some men won't stay to work at any wages when they see the ground," Clarence's brother, Frank Berry, complained.

"Aren't you satisfied?" Frank had asked a man who had decided to quit.

"Yes, I'm satisfied with you, but I won't work for any man in a country where there is dirt like this," he replied. He promptly went up the hillside and began sinking a hole of his own.

All the while, Ethel tried to make a home of their crude, mud-roofed, windowless cabin. At least it was a little better than the tent they had pitched near Ladue's townsite. Ethel might have been a lot lonelier, but the Lippys were living in a cabin just a mile up the valley. Thomas Lippy had already staked on the upper Eldorado, but the timber was better for building a cabin closer to Bonanza Creek. That's where Lippy was when a group of prospectors decided to abandon two of their four claims in order to pursue another creek. Lippy quickly registered No. 16. Like Berry and Stander, Lippy was beginning to hire help. Berry, Lippy, Stander and others already had more gold than they had dared dream of! None of them knew the value of the riches that lay in the dirt heaped up next to the shafts.

Meanwhile, bitter and discouraged after years of fruitless labour, and too late to stake on what was already acknowledged as the richest ground on Earth, Swedish prospector Charley Anderson was drowning his sorrows in Fortymile's saloons. Two disillusioned owners of Eldorado's No. 29 struck up a conversation. Anderson was more drunk than sober, so the two had little trouble in talking the addled miner into signing away pokes worth $800 in exchange for a claim they were secretly sure was worthless. When he woke up the next morning, Anderson floated across the river to ask Inspector Constantine to get his money back from the two men. Anderson's name was plainly visible on the title. Constantine shook his head and apologized: There was nothing he could do. In that case, Anderson figured, he might as well see exactly what it was he had bought. He poled upriver and trudged wearily along the busy creek. Within weeks, his claim proved so rich—on its way to an eventual million-dollar payout—that Anderson was dubbed "The Lucky Swede."

Bonanza and Eldorado Creeks, Yukon
January 1897

Few of the prospectors on the creeks were true thieves. Nevertheless, as winter progressed, tempers flared, voices were raised and fists flew as desperate men became obsessed with protecting wealth they truly thought belonged to them. When the shouting died down, most combatants were eager for an impartial ruling that would determine, once and for all, just who owned what.

Most of the prospectors on the creeks were Americans. Some had been washing gravel in Alaska creeks. Some couldn't read. When they moved to Bonanza and Eldorado Creeks, many had no idea that they were in Canadian territory. Little wonder they were confused by Canadian mining laws. Even in the winter of 1896 to 1897, Canadian claim sizes and the rules of staking were quite clear. The length of a valid claim on a creek was 152 metres. The area of the claim straddled the creek itself to the hilltops on either side of the creek bed. In other areas, it was nine square metres. Each corner of the claim was to be staked by a one-metre piece of wood bearing the date, claim number and prospector's name. All this made little difference to men unfamiliar with another country's mining regulations.

In the chaos of the early days on the creeks, even the most scrupulous prospectors made mistakes. Miners' methods of measurements were sometimes crude, often calculated with nothing more than a length of rope. Through the chaos strode William Ogilvie and his crew. Peering through his precision survey instruments, Ogilvie found many discrepancies. He surveyed Dawson, the townsite Ladue had named after his former employer. Now Ogilvie had to respond to prospectors' pleas to sort out the hastily staked claims in hopes of settling a score of disputes.

Ogilvie was no fool. He knew that moving a stake by a mere foot or two might render one man suddenly destitute and his neighbour instantly wealthy. Yet the government official was not one to shirk his duty. Not only unflinchingly courageous in the face of extreme tension, William Ogilvie displayed another important trait. The surveyor was unassailably honest. He kept himself out of reach of any corrupting influence at a time and place where, quite literally, the most powerful influence of all lay scattered about under his boots.

In reconfiguring one claim just above George Carmack's discovery, Ogilvie found that the owner had taken too much land. The new boundaries created a fraction of pie-shaped, unclaimed land. One prospector on Ogilvie's crew stared at the ground.

"Have you thought of staking it?" he asked Ogilvie.

"I am a government official and not permitted to hold property," was Ogilvie's nonchalant reply. He looked up briefly from his notebook. "You go down if you like and record." The prospector hesitated. If he staked this little piece of ground, he would be ineligible to stake a more valuable piece later. As the day wore on, no other available ground was uncovered. In desperation, he staked the fraction of land and then tried to sell it. At $900, there were no takers. Finally, the prospector decided to work the claim himself. His first shaft proved barren. At the bottom of his second shaft, he exulted. Within a week, he'd hauled up $46,000 worth of gold. Eventually, this sliver of ground (26 metres at its broadest point) earned him more than $400,000! At this moment in history, an ounce of gold was worth a mere $17. If made today, the prospector's "fractional" strike would be worth more than $22 million.

Late one afternoon, Ogilvie was working on Clarence Berry's Eldorado No. 5. Even he was surprised at his findings—No. 5 was close to 13 metres too long. This land no longer belonged to Berry. More bad news: Berry had used up his claim rights, so he had no opportunity to stake it. Even worse, Berry's dump of pay dirt was heaped on the illegally staked piece of land! A year's work and thousands of dollars sitting on top of the ground was in jeopardy. Perhaps even more lay below. Clarence Berry stood by anxiously.

"Let's go to supper," Ogilvie said, turning to the nervous prospector.

"Is there anything wrong, Mr. Ogilvie?" Berry stammered.

"Come out of hearing," Ogilvie muttered as he turned and marched off.

"What's wrong? My God, what's wrong?" Berry whined, stumbling to keep up.

Striding toward the Berrys' cabin, Ogilvie revealed his findings.

"What'll I do?" Berry despaired.

"It is not my place to advise you," Ogilvie stated calmly. Then, after a pause, he asked, "Haven't you a friend you can trust?"

"Trust—how?" Clarence asked desperately.

"Why," Ogilvie explained matter-of-factly, "to stake that fraction tonight and transfer it to yourself and your partner."

Berry showed Ogilvie inside, asked Ethel to fix him supper, and dashed about seven kilometres up the creek to find his trustworthy companion. Later that night, the two prospectors sat in the cabin and quizzed Ogilvie about the proper way to stake a fraction. Then, hidden by darkness, Berry's friend hammered in his stakes. In return for the claim transfer, Berry ceded him an equal length off the lower end of the property. The following spring, that fraction yielded $160,000 worth of gold.

CHAPTER

6

A Secret
Locked in Ice

IN SHARP CONTRAST TO CONDITIONS on the rest of the con-
tinent, a handful of men had become enormously wealthy
overnight in a remote, snowbound territory few Canadians
and even fewer Americans had ever heard of. On small, unknown frozen creeks, fortunes were being found. Elsewhere
in Canada and the United States, fortunes were lost.

In the era ironically known as the Gay Nineties, thousands became impoverished as the United States and most
parts of Canada were held in the merciless grip of economic
depression. Railroads had collapsed. In their wake, banks,
mortgage companies and corporations had failed. Men
died or were thrown into prison as a result of violent labour
strikes. The ranks of the jobless swelled, and many lived

close to starvation. People from coast to coast were desperate for relief, or at least were hoping against hope that things might be better someday, somehow, somewhere.

Fortymile, Yukon
January 1897

William Ogilvie was pleased. Captain William Moore's appearance answered the question of how the isolated surveyor was going to inform the Canadian government of the area's enormous changes. Delivering the US mail the 1,120 kilometres from Fort Constantine to Juneau, Moore could put Ogilvie's letter on the steamer to Vancouver. From there it would be carried by train to Ottawa.

Moore's energy was astounding. A decade before, Ogilvie had doubted the old man's stamina. Now, Moore was vigorously mushing his dogs along in sub-zero weather, while most gentlemen in their seventies wanted nothing more than a snifter of brandy by the fire.

For his own part, Moore was not surprised that Ogilvie was still at Fortymile. As Moore had made his way to St. Michael months before, a three-day snowstorm had dashed Ogilvie's plans to leave the Yukon. As Moore reached Circle City, Alaska, the town had become completely iced in. Since he could travel no farther north, he had made a profitable decision. He'd decided to carry a sackful of mail back up the frozen Yukon River. Swaddled in furs and woollens, Moore was on the trail again, with Ogilvie's letter to the Canadian government in tow.

A week before, three hardy young mushers had set off on a similar journey. They left Fortymile in an attempt to set a new record to the coast. Within a few days, Moore had caught up to the trio, now starving and exhausted, and offered them a life-saving lift in his dogsled. Moore's journey did not end at Juneau; he was goaded on by the knowledge that few on the outside knew of the big strikes. He desperately wanted to give the news to his associates in Victoria.

Weeks later, Ogilvie's letter arrived in Ottawa, where it was duly filed and forgotten.

* * *

In the log town of Circle City, Alaska, which William Moore had left behind, men were becoming increasingly restive. Moore's story of Klondike strikes had been shrugged off as rumour and exaggeration. Then two traders arrived with samples of so-called Eldorado gold. They revealed letters from men who had purportedly become wealthy overnight.

"This is one of the richest strikes in the world," the bar owner read aloud to the 75 silent men crowded around his bar. "I myself saw $150 panned out of one pan of dirt, and I think they are getting as high as $1,000."

Perhaps not exactly $1,000, though. Ironically, the bald truth was so fantastic that few would have believed it. In the biggest settlement in this part of the frozen north, life went on as usual. Then, Circle City saloon operators received a letter from their distant partner in Fortymile—Bill McPhee. Like McPhee, the Circle City saloon operators were veteran "sourdoughs," who shared years of prospecting experience and a healthy skepticism for tales of gold. Yet, what McPhee wrote was intriguing enough that they mushed more than 320 kilometres upriver to take a look. It was all true, they reported by mail. Interest increased.

Now, a month later, the latest batch of Klondike mail was delivered to another saloon in Circle City.

"Gimme some beef tea!" the cold, exhausted mail courier demanded. The saloon owner ignored him, pawing through the bag of letters. Men began to gather around. Just a few hours before, another courier had arrived from upriver, spreading the Klondike's tale. Because the courier had no proof, the response was either laughter or anger. Letters might contain proof, and they had finally arrived. The saloon owner ripped one open. A few seconds of silence ensued as he devoured its contents.

"Boys, they're right!" he shouted, vaulting over the bar. "Help yourself to the whole shooting match!" he told his stunned customers. "I'm off to the Klondike!"

The exhausted mail courier never did get his beef tea. After enduring a few minutes of the raucous free-for-all inside the saloon, he sought escape from the laughter and shouts, making his way toward a friend's cabin for something to eat. A mob followed, pestering him with questions.

Within 48 hours of the saloon owner's defection, Circle

City was all but deserted. There would be no more waltzes at the dance hall, no more concerts at the opera house. The saloon operators were all gone, along with most of the prospectors. The prostitutes and the dance-hall girls would soon follow. At first, Jack McQuesten decided to stay, selling his supplies to those who remained. Soon, however—heedless of the NWMP's customs duties—McQuesten left his waterfront store behind. He too was off to the Klondike, along with what seemed to be everyone else living up north. In reality, there were fewer than 1,500 prospectors at the Bonanza and Eldorado Creeks. The registered claims at the end of November numbered less than 350.

Bonanza and Eldorado Creeks, Yukon
Winter 1897

At a time of year when men stayed close to their stoves, there was a steady flow of traffic up the Yukon River. They trudged past the deserted cabins at Fortymile, through the doors of John Healy's store at Fort Cudahy and into the tent city of Dawson. Only dreams of sudden wealth could rouse people to venture forth in air that was so cold it turned a man's breath to icy crystals in front of his face.

On the creeks, shadowy figures muffled against the elements continued to burrow like moles. They dug below the snow-topped mounds of black earth that marked each claim. Day after day, below the surface, the lonely, monotonous burn-and-dig routine continued as prospectors

Hand on hip, looking every inch the debonair man, Skookum Jim poses at No. 1 on Bonanza Creek, 1899.

built fires to thaw the frozen earth enough to shovel it into buckets. The buckets were then hauled to the snow-covered ground above.

Every now and then, someone would decide that enough was enough. It was easy to sell to a legion of latecomers such as Circle City resident Jack McQuesten. His purchase yielded $10,000 in gold. McQuesten's two travelling companions purchased a half interest in another claim. Before they could do much more than lower themselves into the shaft, they accepted an offer of $20,000. The money quickly changed hands for an interest

in yet another claim. They sold the first half interest for a
$10,000 profit! There were quicker, easier and cleaner ways
to make a fortune than prospecting.

Buyers often prospered at the expense of sellers. That
winter, one prospector pocketed $3,000 for his claim, not
knowing that he had practically given away one of the richest claims on the creek. Another man sold half of his claim
for $800. Later, that land was valued at over $1 million.

The biggest buyer of all was Nova Scotian Alex McDonald.
By mid-winter, the house of "claim" cards that McDonald
had constructed with buy-and-sell tactics was as unstable
as it was complex. "I've invested my whole fortune," he admitted without shame. "I've run into debts of $150,000," he
said, adding confidently, "but I can dig out $150,000 any
time I need it."

By today's standards, these amounts are still
significant. More than a century ago, they were almost
beyond imagining. Five thousand 1897 dollars had the
same value as $100,000 today. And $5,000 was exactly
what Alex McDonald needed to borrow. Ron Crawford, a
former Seattle court clerk, was making money drawing up
prospectors' legal documents. McDonald decided to tap
him for a loan. When Crawford asked what interest he was
prepared to pay, McDonald shook his head.

"Interest is always working against you," McDonald explained. "I can't sleep at night when I think of that." Instead,
McDonald offered 30 metres of No. 6 below the Bonanza

discovery, at 50 percent. As security, he then offered 35 percent of 11 metres of No. 27 Eldorado, along with a mortgage on No. 30 Eldorado.

The man adopted McDonald's own maddening habit of making people wait for a decision. He told him he would think about the offer. Crawford did more than think. He raced off to persuade a saloon owner to part with $5,000 on a promise of a half interest in the mortgage on No. 30 Eldorado. After the loan was made, the former bank clerk flipped part of No. 27 Eldorado to a Dawson barber for $5,000. The barber put down his razor, picked up a shovel and dug up $40,000 worth of gold. And the outside still remained unaware of it all.

That winter, the enormous wealth that lay in the ground and in the dirt piled up on top of it was a secret known to a handful of individuals. By early spring, the secret had been leaked. The *Vancouver Daily News Advertiser* carried a story in February. Another appeared in a Chicago newspaper in March. Few noticed; even fewer cared. Thus the secret remained locked away until spring released winter's icy grip on the rivers.

7

The Treasure Ships

TWO TINY STEAMERS, THE Alaska Commercial Company's *Alice* and the NAT&T's *Portus B. Weare*, chugged down the Yukon River toward the Bering Sea and the distant coastal city of St. Michael, Alaska. On board the vessels were Ethel and Clarence Berry, Salome and Thomas Lippy, Jack McQuesten, Joe Ladue, William Ogilvie and others. Passengers included former NWMP officers, clerks, lawyers, laundrymen, carpenters and cowboys. Their original homes had been in eastern Canada, a dozen US states, Scandinavia, the United Kingdom and continental Europe. For all their differences in background and language, they had much in common. It had been at least one year since they had seen civilization. For some, the time away from cities and towns had been even longer. None of the passengers were the same as they had been when they had first arrived in the North. And now, every one of them was going to the outside.

Each carried clothes and personal items, which had become relatively meaningless—for each carried another far more important item. It was packed inside the gangway in suitcases, blankets and bedrolls and also in boxes, belts and leather bags. It was awkward luggage, for collectively it weighed tons. It was gold. Individuals in this small, yet diverse group of travellers were now wealthy beyond their wildest dreams.

When they docked at St. Michael, Alaska, their wealth was momentarily forgotten. Here were real onions and turnips! Tins of pineapple and cherries! It had been a long, long time of little more than beans, bacon and flapjacks. The wealthy passengers were offered new choices about what kind of life to lead and how and where to live it. Life would never be the same for them. Still, once they reached their northern destination, life would never be the same for tens of thousands of other ordinary people, as well.

San Francisco, California
July 15, 1897

It was a typical day on the wharf near the foot of San Francisco's Market Street. The regular work gangs were loading and unloading the ships. As always, passengers were waving down to little groups of well-wishers standing on the docks. Nobody paid much attention to the squat, grimy little vessel that edged toward the wharf. The *Excelsior* was just another weathered wooden coastal steamer docking after its nine-day voyage from Alaska.

A few minutes later, some people on the dock took interest in the grubby men walking down the *Excelsior*'s gangway. Shabby and bearded, prospectors were a strange bunch. This time, there were women in the group, as shabby-looking as the men. More passengers stumbled down the gangway. Salome and Thomas Lippy struggled to maintain their balance, a small suitcase swaying precariously between them. Workmen on the wharf paused and glanced at one other. It took *two* people to carry down a *suitcase*? Men reined in their horse teams and walked over to take a closer look. Soon, the small gathering of onlookers attracted others. Something was going on.

Mild curiosity quickly turned to jaw-dropping amazement. The bags and boxes were heavy because they were full of gold! Within minutes, the passengers were forced to elbow their weighty baggage through the growing crowd. One *Excelsior* passenger stepped in front of a four-horse team. Where was the mint? The mint was closed, someone replied. Where was the nearest smelter? Montgomery Street, someone shouted back. A driver was hired on the spot, and the group threw their gold and themselves into the wagon and off they went, the excited crowd hurrying to keep up. At the Selby Smelting Works, a spellbound audience watched as the precious contents of jars, bags and sacks were poured onto the counter "like a pile of yellow shelled corn," according to one awestruck witness.

Newspaper reporters wanted specific figures. The gold in the Lippy's suitcase weighed close to 91 kilograms. At $17 an ounce (about 31 grams), they had carried more than $54,000 down the gangway! Another passenger had carried $35,000, and a former Seattle laundryman, $15,000. His riches seem paltry compared to some others, yet at a time when $700 was a typical annual salary, the lowly laundryman was likely set for life. Turn-of-the-century economic realities made the hauls breathtaking.

Up in Calgary, the *Calgary Herald* chased down a NWMP sergeant who was happy to tell the reporter about former Mounties who had augmented their meagre pay in spectacular fashion. One Mountie had hiked up the creek from his post one Sunday, staked a claim and was back in time for Monday's 7 a.m. roll call. He didn't have to work the claim; he simply sold it, pocketing $40,000. It seemed that Superintendent Charles Constantine granted leave quite routinely on the assumption that a wealthy policeman was a happy policeman.

Joe Ladue towered above them all. Without ever swinging a pick or dipping a pan, he held paper worth $5 million.

Excelsior's agents had to close their wickets. There were no more tickets to be had. The return trip to Alaska had quickly sold out. The desperate begged, pleaded and even waved money, but to no avail. Ticket holders foolish enough to brag or luckless enough to loiter about the agents' office were tripped up, dragged into alleys and pummelled, their pockets searched and precious tickets stolen.

San Francisco telegraphers quickly tapped out the news to points east, south and north, as the *Portland*, with more than a ton of gold in the hold, beat its way past Vancouver Island to Seattle. When the steamer nudged the dock, the city was ready and waiting.

Seattle, Washington
July 1897

LATEST NEWS FROM THE KLONDIKE stated the large headline on the front page of the *Seattle Post-Intelligencer*. Then, just below it, the page blurted the words GOLD, GOLD, GOLD, GOLD! as if the newspaper was unable to hold back any longer and had succumbed to the peculiar malady that had already stricken thousands of its readers.

Even before the *Portland* sailed into view, the Seattle waterfront was swarming with men and women suffering from symptoms of what would prove to be a highly contagious disease—fits of shrill laughter, wide-eyed babbling and irrational urges to abandon work, home and the city.

"Show us the gold!" some shouted to prospectors who lined the ship's rail above the throng. The men happily obliged, waving caribou-hide pokes about. The crowd went mad. In that crowd was one prospector's destitute wife, who watched her husband haul $112,500 worth of gold off the boat. Another wired his impoverished washerwoman wife and told her to stop scrubbing. He had brought down $50,000. Mere hours after the Berrys and more than 60 other prospectors stepped off the gangway, gold fever had become an epidemic.

Streetcars were jammed as conductors and operators came down with gold fever. Becoming quickly infected by the stories they wrote, most *Seattle Times* reporters left the newsroom forever. In downtown stores and offices, scores of clerks were overcome by the swift-moving virus. Fire halls and police stations were also hard hit. The contagion swept up the steps of city hall, reaching into the mayor's office. Attending an unexpectedly timely San Francisco convention two days earlier, Seattle's mayor became one of gold fever's earliest victims. He simply wired his resignation. Buying a ticket north wasn't good enough for him, either. He decided to buy an entire steamer to take him and paying passengers to the land of the midnight sun. The mayor was on to something. Within days, every regularly scheduled sailing was booked solid for the remainder of the year.

After the isolation of the North, many arrivals couldn't stand the attention. Some fled as far as Chicago and New York. Few found peace anywhere. Given the circumstances, the Berrys' choice of destination—San Francisco—was a questionable one, but at least the Lippys were there. Once settled in the Grand Hotel, the Berrys relented and began granting interviews.

"Two million dollars has been taken from the Klondike region in less than five months," Clarence Berry told the *San Francisco Chronicle* reporter, calling the area, "the richest gold field in the world." He offered proof: a collection of vials and bottles filled with gold, just a mere fraction of the $130,000 he had carried down from the Klondike.

Reporters were fascinated that ladies would venture into the Yukon. Salome Lippy and Ethel Berry sat down with reporters from the *Chronicle*, the *Sutter Creek Record* and other journals to answer questions. One author asked what they did for amusement so far from civilization. The women chuckled.

"We did not think of anything but sleep and rest," Ethel explained. "That was the main reason we didn't die of homesickness. We had no time to think!"

"Nobody bothered much about amusements," Salome agreed. "Everyone was busy and kept busy all the time. I did my work!" she added, somewhat defensively. "Mining is genuine toil. When Mr. Lippy finished, he wanted to rest."

Given the hard work and rugged, primitive conditions, puzzled reporters asked why the ladies had gone up to the Yukon in the first place.

"I went because my husband went, and I wanted to be with him," Ethel replied simply.

What advice would Mrs. Berry give to those thinking of joining the stampede?

Ethel levelled a look at the reporter. "Why, stay away, of course! It's no place for a woman." Then she reconsidered. It was no place for a single woman, she clarified.

* * *

Studying the pandemonium raging around him, Soapy Smith smiled with satisfaction. As inured as he was to displays of uncontrolled emotion, it was all he could do to rein himself in. Many of the people passing on the street were so distracted that a man could lift their wallets from their pants and half of them wouldn't notice. None of them wanted to be in Seattle. They all wanted to be in the Klondike. Where they went, Smith had already decided, he would follow. He wouldn't go alone, either. He had carefully selected members of his Denver crew to assist him in taking full advantage of the situation. This time it was gold, not silver, that brought the gullible together, but the reason scarcely mattered.

Soapy Smith never sank a mine shaft or swirled gravel in a pan. He hadn't dirtied his hands since he helped drive longhorns up the Chisholm Trail years before. When he'd been in Colorado, Smith used a deck of cards and more creative schemes to part the gullible silver miners from their money. Once, dozens lay down $5 each, betting that a particular cake of shaving soap purchased from Smith hid an elusive $20 bill inside the wrapper. Of course, it did not. From then on, he was "Soapy" Smith.

In a scene repeated in every major West Coast port, the *Roanoke* docks in Seattle during the frenzied Klondike summer of 1897.

His gambling house was really an educational institution, he once told a jury when two victims brought charges. He had even warned his customers they had no chance of winning against him. There it was, nicely scripted in Latin for all to see: "Let the Buyer Beware." The two victims, he reasoned, would never gamble again.

By this time, Soapy Smith's ambitions had long since exceeded his one-man con games. The lack of formal law and order made Colorado a bandit's paradise. Soapy simply took over most of the towns in which he operated. He hired deceptively genteel accomplices to bring the suckers in and others to take their money, sometimes, when more subtle means failed, at gunpoint or after a few convincing blows from a sap. Smith was betting that the Klondike would be bigger and better than Colorado. As everyone knew, Soapy always won a bet.

Soapy Smith felt a tug on his coat sleeve, turned and looked into a face he hadn't seen in ages—it was a policeman from his Colorado days. He knew Soapy Smith well enough not to be overly surprised to see him on the West Coast at this particular moment. The two men began to talk about what everyone else was talking about.

"I'm going to be the boss of Skagway," Soapy Smith confided casually. The policeman nodded knowingly and folded his arms, waiting to hear more. "I know exactly how to do it," Smith continued. Then he had a sudden thought. "If you come along, I'll make you chief of police," he offered magnanimously.

"A team of *mules* couldn't drag me to Alaska!" replied the policeman, seemingly immune to the infection that was ravaging the entire coastline. Soapy Smith shared this immunity to gold fever but had decided to head for Alaska anyway. He suffered from a compulsion of a different sort.

In truth, Soapy Smith didn't yet know if his final destination was Skagway, Wrangell, Juneau or perhaps that new town near the gold fields, Dawson City. Smith would take a methodical approach and visit them all. He knew what he was looking for, and he would recognize it the instant he saw it.

Arranging passage up wouldn't be too difficult. His "associates" would get steamer tickets for them one way or another.

* * *

Gold fever spread north across the border. Patrons in Vancouver's bars were soon singing up a storm:

> Klondike! Klondike!
> Label your luggage for Klondike
> For there ain't no luck in the town today,
> There ain't no work down Moodyville way,
> Pack up your traps and be off, I say,
> Off and away to the Klondike.
>
> Oh, they scratches the earth and it tumbles out
> More than your hands can hold;
> For the hill above and plains beneath
> Are cracking and busting with gold.

Within days of the arrival of the treasure ships, so many Vancouver deckhands came down with gold fever that loading boats for scheduled trips to Vancouver Island was a challenge. For the Canadian Pacific Navigation Company (CPN) and Union Steamship Company, however, a labour shortage was a small price to pay for a sudden end to the lingering waterfront depression. Three days after the *Portland* docked in Seattle, Vancouver's vanguard contingent of prospectors was sailing out of Burrard Inlet on CPN's *Alki*. Two days later, the Union's *Capilano* set sail for Dyea Inlet, loaded with horses, cattle and drivers. Crowded passengers slept on hastily hammered berths and deck bunks.

When the fever hit Victoria, CPN's twin-funnelled *Islander* was so overbooked with frantic Americans unable to find passage at US ports that the smaller *Tees* was pressed into service. Victoria hadn't seen anything like this since the wild days of the Cariboo Gold Rush. Outside the customs building, the line of Yukon-bound Americans straggled far down Wharf Street.

Eastern journals were quick to report gold's arrival and the public's reaction to it. SEATTLE HAS GONE STARK, STARING MAD ON GOLD wrote the *New York Herald*. COAST AGAIN GOLD CRAZY reported journalist Tappan Adney. Reporters found the simple, evocative word "gold" no longer adequate. Enticing tales were written about "hard, solid" gold, "rich, yellow" gold and "shining" gold, further firing the imaginations of susceptible readers. In North America, gold's lustre had surpassed its legendary lure years before.

Gold's scarcity in the United States had become a national concern, as population growth outpaced production. Many began to horde the yellow metal in drawers and under mattresses, unwittingly worsening their economic fates. As railroads went under and mortgage companies and banks went broke, folks began stashing coins and paper money as well. Klondike gold represented an incredible once-in-a-lifetime opportunity. Many had little to lose by taking a chance. To finance the dream, tens of thousands began to open drawers and lift mattresses.

The world needed to know everything it could about the Klondike. Toward the end of summer, journalist Tappan Adney received a telegram from England's *Harper's Weekly* and the *London Chronicle*, asking him to get to the Yukon. Two thousand New Yorkers were planning to do the same thing. Journalism had its privileges, and two days later, at the Canadian Pacific Railway's (CPR) New York office, the reporter managed to buy a New York to Dyea Inlet ticket, including steamer passage from Victoria. Every place he went, Adney talked with Klondike hopefuls. In Chicago, he interviewed John J. Healy's financier, Portus B. Weare.

"I'm afraid there's going to be trouble in the Klondike country soon," Weare prophesied, telling the surprised journalist that both the NAT&T and Alaska Commercial Company had been working hard to get sufficient supplies into the Yukon before winter. Time, distance and the huge influx of men were working against them, he lamented. The company couldn't possibly feed everyone. He painted a gruesome picture.

"I greatly fear that a vast and uncontrollable multitude will rush up to Juneau all through September and October and invade the country in a mad quest for gold," he told Adney. "There will be thousands without proper food and equipment, and they will plunge into the midst of the snow and terrors of an Arctic winter ... there can be but one sequel." Healy's financier urged everyone to carry over enough provisions to feed themselves.

Tappan Adney planned to join those climbing the White Pass. In Winnipeg, where he stopped over to purchase supplies, he was stunned to learn that there were no fur coats left anywhere, and many other cold-weather necessities had already been snapped up.

"All aboard for Minneapolis, Seattle and Klondike!" the Chicago depot master called out. Reporter Edmond Hazard Wells stepped up onto the Great Northern on the first leg of his journey to the Yukon. Wells was on assignment for a newspaper syndicate that included the *Cincinnati Post*. E.H. Wells was the ideal man for the job. He had been to the Yukon before. In 1890, as a member of an expedition sponsored by *Frank Leslie's Illustrated Magazine,* Wells had been among the first white men granted the privilege of climbing the Tlingit's closely guarded Chilkat trading route. He and others had floated down the Yukon River and had seen prospectors at work on Fortymile's creeks. Now he wrote his colourful dispatches with particular zeal. For some, gold fever was a chronic illness. Wells had been a "carrier" for years. Now, his fever had broken out again.

8

Journey into Hell

NORTH-WEST MOUNTED POLICE superintendent Charles Constantine's vision of thousands of possibly lawless men overrunning the Yukon was becoming a reality. It would be months before Constantine, George Carmack, Skookum Jim and some of the other early claim holders witnessed the full calamity. However, two other men, an old visionary and his son, experienced a frightening preview of what was to take place later on the Yukon River.

Skagway, Alaska
July 1897
William Moore had toiled 10 years to wrest a 65-hectare townsite out of the tangle of coastal forest in Skagway. Now

it was done. The site was staked out, the sawmill was up and running, roads—complete with bridges—ran in and out of the forest, and much of the first wharf was complete, as well as two neat log houses to house the work crews who were building it. Moore's entire life had seemed a prelude to this moment, a moment he had dreamed, schemed and fought for.

Two years before, in Victoria, he had met with the representative of a British syndicate to argue that his pass—the White Pass, not the Chilkoot Pass—was the right route for a wagon road or railway. It had been a passionate, persuasive performance. Moore had been rewarded with men and money and in return had granted the syndicate prime interest in his wharf and property. It was a paltry price to pay for the opportunity to realize his vision of creating his own town and Yukon roadway. Now, Moore and his son waited eagerly to greet the people who would make them rich.

Moore watched the steamship float silently up the inlet. The first sound that reached his ears was the ship's whistle, echoing off the forested slopes that surrounded the bay. The second sound that reached his ears was the frightening cacophony of frenzied human beings. There were hundreds of them—screaming, shouting and cursing.

Moore stood rooted to the spot in disbelief. Who was in charge? Who should he meet? His questions went unanswered. Before Moore knew it, men with axes were cutting down *his* trees, and men with tents were pulling out *his* stakes! Crazed horses raced pell-mell across the beach

A few of the stampeders who pushed aside Captain William Moore at Skagway stand dwarfed by their mountains of supplies.

and up the wooded riverbank, handlers in hot pursuit. In dismay, William Moore realized there was no one in charge, no one with whom to discuss purchases, leases or rights.

"Trespassers," Moore screamed, but the puzzled, frantic strangers simply shoved him aside. And still they came. In the days that followed, ship after ship delivered the gold-hungry hordes. Soon, the rasp of saws and the blows of hammers were added to the waterfront din. Rude buildings were thrust up on either side of a muddy trail. Somewhere, above the cries, shouts, pleas and laughter, the sound of a piano floated.

A meeting was held to "set up the town." A committee was struck and a plan of blocks and lots proposed. Nobody consulted William Moore. Instead, a bartender, Frank H. Reid, was appointed surveyor, perhaps for no other reason than he already possessed the instruments, traded in by some would-be miner for a good time at the bar. Reid decided to call himself city engineer.

Reid met with the Moores, not to ask for advice, but to suggest that they stake out a few acres for themselves before they were completely overwhelmed. Upon completing his surprisingly competent survey, Reid realized they had a problem. The wharf bunkhouse, which Moore and his wife used as a home, occupied land destined to be a downtown intersection.

The committee's crewmen reported that when they approached the bunkhouse with their equipment, the old man had gone into a rage. He had been there long before they had, the apoplectic Moore screamed. While his weeping wife watched, Moore yelled that he had no intention of moving for anybody. The committee gave Moore 24 hours to vacate the bunkhouse. Moore slammed the door in their faces.

Two days later, a mob armed with saws and sledgehammers approached the home. Snatching up a steel bar, Moore jerked open the door and charged the crowd, swinging as he went. A man stepped forward and brought his axe down on the front door. Moore whirled around and brought his bar down on the man's arm. The man yelped in pain, and the mob fled. Captain William Moore had won the battle, but he had already lost the war. The Moores were soon lodged in a makeshift hotel, and his bunkhouse was moved near the waterfront to house the syndicate's crew. Work on the wharf continued. As even Moore realized, there were bigger things at stake than one's personal property.

White Pass, Alaska
August 1897

In the middle of the Skagway chaos, journalist Tappan Adney rarely stopped taking notes, and he never stopped asking questions. A California mining engineer surveyed the chaotic Skagway waterfront. "I have never seen people act as they do here," he admitted to Adney. "They have lost their heads and their senses. They have no more idea of what they are going to do than that horse has," he said. Adney concurred. He'd brought a train of eight packhorses himself and watched in disbelief as other men who'd never handled horses tried to tie on packsaddles. The "diamond hitch" was the knot required to keep the packs tied securely for hours of jostling over rough terrain. Adney was a practised packer and knew the knot well, but few others had heard of it, and no one seemed to know how to throw it. Worse still, many of the horses were nothing more than "ambulating boneyards, the infirm and decrepit" with "dropping heads, listless tails," he reported.

Adney had encountered this kind of ignorance days

before on the Victoria wharf. One gentleman was leading a pack of dogs—Irish setters—and had taken along a lawn-tennis set. He was off to Alaska, "just for a jolly good time, you know."

A fit and agile outdoorsman, Adney was keen to hear more about the White Pass he was about to climb. He was curious that so many were already coming back down.

"The road is good for four or five miles—it's a regular cinch," recounted one disillusioned, defeated stampeder. "After that, all hell begins."

The problem, said another, was "the inexperience of those who are trying to get over. They come from desks and counters; they have never packed, and they are not even accustomed to hard labour."

Tappan Adney's fellow journalist, Edmond Hazard Wells, concluded that the would-be prospectors would literally die of ignorance. "Thousands of corpses will lie on the mountainsides and in the valleys," the *Cincinnati Post* reporter predicted. However, Wells soon discovered lack of know-how was not the only reason for agony on the White Pass trail. In fact the trail did not exist.

Wells was no stranger to the Yukon. He had explored the region in 1890 and was among the first white men to climb the Tlingit's closely guarded Chilkat Route over the mountains. His past experience paled in comparison to this new one. Initially swept away by the drama he was witnessing, Wells provided his readers with positive, exciting stories of "800 men at work bridging chasms" and "gulches . . . resounding to the ring of a thousand axes."

Just two days later, after labouring with a tiny sweat-streaked gang of fewer than 200 men, a disillusioned Wells wrote, "The gold hunger grew so keen that after a few corduroy bridges had been thrown across marshes and streams, the attempt was abandoned and a general helter-skelter ensued." It was, the reporter confessed, "every man for himself."

A few miles up, Adney was already "ankle deep in sloppy, slimy chocolate-covered mud." Lack of planning and coordinated effort, and most of all, lack of time, would have horrific consequences. Adney was there to witness the first of them.

On a rocky slope, an empty packsaddle hinted at an accident. Adney and climbers needed no one to tell them what had happened: a horse had rolled down off the trail and fallen hundreds of feet below. A few minutes later, they came upon three dead horses, "two of them half-buried in the black quagmire." The stench was unbearable. Abandoned by uncaring owners, scores of spent, crippled and bloodied animals stood motionless, waiting for death to end their agony.

Adney talked to one man who claimed to have seen a horse deliberately walk over the face of Porcupine Hill. "It looked to me, sir, like suicide," the man stated matter-of-factly. "I believe a horse will commit suicide, and this is enough to make it." He shook his head and gave Adney a haunted look. "I don't know, but I'd rather commit suicide, too, than be driven by some of the men on this trail."

Slow-burning, wet wood fires sent dense clouds of smoke and the pungent smell of roasting horseflesh wafting down the trail. After two days, knowing it would only get much worse, Tappan Adney joined a legion of others forced to turn back to Skagway. His goods were waiting at Dyea Inlet with his companions.

As Adney neared Chilkoot's infamous Scales (formerly the site of large scales used by Native packers), he had a visitor at his tent, "a wreched, thin white cayuse . . . thin as snakes and starving to death." The horse had made its way down from the pass when abandoned by its owner. The horse put as much of its body into the tent as it could, just to get out of the rain and feel the warmth of the stove.

The next day, the horse was still there. Adney pulled out his .44 revolver, walked slowly up to the animal and mercifully put a bullet behind its ear. Before long, the carcasses of horses lay all about.

Hazard Wells struggled up White Pass, now jammed with goods, horses and "men with haggard faces, unkempt hair and disordered attire . . . plodding along, groaning under the weight of 38 kilogram packs. Perspiration . . . bursting from every pore and their breaths coming hard and fast."

On August 15, Wells and "about one dozen men" had reached the summit and staggered on to Lake Bennett. Wells calculated they had left nearly 2,500 others slipping and stumbling behind them and estimated that a mere two dozen would make it. Hundreds were abandoning supplies

Heartless cruelty in Dead Horse Gulch. Thousands of animals perished on the White Pass route.

and retreating back down to the coast. Exhausted and traumatized, Wells wrote, "I wish now, without loss of time, to send back word of warning to everybody. Keep away from Skagway and White Pass."

The agony and anguish on the trails continued long after Adney, Wells and others had reached the summit. To urge his exhausted oxen forward, one maddened prospector built a fire under the animals, slowly roasting them alive.

"Clear the trail! Man dying!" was the cry of those desperate to try and save the life of an ill or injured friend. Still, hundreds would scarcely glance at the dead bodies of gold-seekers. During that hellish winter of 1897 to 1898, more than 2,000 animals and an uncountable number of men perished on the trails and in the Yukon's frigid waters.

Yukon River
October 1897

For the previous two or three years, ice had locked in the Klondike River by mid-October. By October 12, Tappan Adney and his party were entering Lake Marsh in their newly built boats. They were still hundreds of kilometres from their Dawson City destination. The journalist realized it was now a race against time.

Caught in the turbulent waters of the dreaded Miles Canyon, Adney's Dyea Inlet companion, "the imperturbable" Al Brown, earned the journalist's respect and admiration for his navigation skills. The two men raced breathlessly between the canyon's perpendicular basalt walls. White Horse Rapids loomed ahead. As they manoeuvred through the rapids, water drenched the two, filling their boat. Then "a big side-wave [took] the little craft, [spinning] her like a top." Suddenly, the two found themselves floating in a quiet eddy. Later that night, as they dried out at a sandy cove, Adney asked Brown if he had been scared.

"Why, no," Brown replied thoughtfully. "You said it was all right. I suppose you know," he shrugged. "It's your boat and your outfit." Adney couldn't have asked for a better partner. "I believe," Adney wrote, "that if a charge of dynamite were to explode under Brown, he would not wink an eyelash."

A few nights later found them nearing the halfway point on the river journey to Dawson, at Fort Selkirk. The manager gave the men a less than enthusiastic welcome. "So many are coming in unprepared, either without outfits or common sense," he complained that night, over a rare bottle of scotch. "They ask me what the price of flour is in Dawson. I tell them it has no price. People on the outside talk as if the steamers on this river run on a schedule," he scoffed. "There would be winter shortages," the manager warned. He looked hard at Adney and jerked his head in the direction of the nearby cabin where his companion was sleeping. "Has the young man with you a sufficient outfit?"

"No," Adney admitted, shaking his head, adding somewhat defensively, "But he's strong. He's willing to take the risk of finding something to do and buying grub."

"Very foolish thing for him to do," replied the manager, staring into his half-empty glass. "Many people are short and more may have to leave before spring." He looked back up at Adney, and his eyes narrowed. "Time was when it would go hard with a man who was responsible for bringing in a person like that."

The man was a crank, Adney thought later. But then, "every man becomes a crank who stays long in this country," he concluded.

The next morning, fog delayed their departure. Standing

around the trading post's big stove, Adney and Brown fondled the skins and furs. The manager wanted them to take some moosehide with them to Dawson. Moosehide was in great demand for moccasins, he explained. After setting some aside, he hesitated and then shoved them back in the pile. "No," he said. "I won't send them with you. I don't think you'll reach Dawson."

* * *

Adney and Brown were on the river again, watching ice chunks flow past, some as big as wagon wheels. Again and again, they were forced to pound the ice off the oars. Mitts froze stiff; moustaches became a mass of icicles. No matter how hard they worked, they could not keep warm. That night at camp, after a hasty dinner of flapjacks, beans and hot tea, they drifted off to sleep to the ominous sound of ice moving downstream in the dark.

Their eyes snapped open at a sudden, long roar from the bank. The men dashed over to the water's edge. A 12-metre-long ice floe was tugging menacingly at the boat. The desperate men yanked their supplies to safety and shoved the empty boat beyond the reach of the river's ice.

Shoulders heaving from the effort, his head clouded in vapour, Adney watched Brown stumble back toward the fire. He turned and looked back at the boat, its graceful contours barely discernible in the gloom. That little wooden boat was the only thing that stood between them and certain death.

Adney looked up toward the clear heavens and watched the northern lights' pale green patterns dance across the sky. The country was beautiful. He peered over in the direction of the groaning ice. It was beautiful, all right, beautiful but deadly.

Lying on the cold, hard ground, Adney reached for his notebook. "We have abandoned all hope of reaching Dawson," he wrote. He looked into the fire and bent again to his book. "There are hundreds on the river this night feeling as we do."

The next morning, the river was filled with grating, grinding ice floes. Drifting through the mist, surrounded by bleak, grey-black mountains, their tall spruce whitened by frost, the men in their little boat were, Adney wrote, "a picture of loneliness." They spied two men struggling along on the ice near the shore. "There is no grub at Dawson," one cried. "If you haven't an outfit, for God's sake turn right back where you are!"

Two days later, as Adney and Brown moved down the swiftly flowing river, they came upon a group that had just pulled their boats onto shore. One man in the group shouted a warning: just below, the ice had jammed the mouth of the White River. Adney knew there was just one thing the two of them could do, and they had to be quick. They steered their craft as close to the shore ice as they could, planning to drop into the safety of a quiet eddy. With a start, they realized that they were trapped in the swiftly moving ice floes. They watched helplessly as the calm eddy receded into the distance. The ice began to grind menacingly against the prow of the boat.

Suddenly, a large chunk of ice crashed into the bow. Gasping with horror, the two men expected to hear the frightening sound of splintering timbers any second. They tensed for the desperate jump onto the bobbing ice. Instead, the impact of the ice floe gently tilted the boat up and the large chunk of ice rotated harmlessly away.

Moving along, Adney and Brown stopped to talk to others who had hauled their boats ashore. For some, bound for the tributaries of the Stewart River, their Yukon River journey was over. Some had stopped out of fear, still others were halted by disquieting news of food shortages and thievery in Dawson. It was getting so bad that two men had been shot for breaking into caches. In time, the two travellers witnessed for themselves just how tense the situation in Dawson had become. It was snowing when Adney and Brown rowed out into the river the next day. Ice soon covered their oars, gunwales, boxes and bags.

Days later, just two kilometres from their overnight camp, the two men saw a large number of boats, tents and people on the bank ahead.

"How far is it to Dawson?" Tappan called out.

A man on the shore turned and cupped his hand to his mouth. "This *is* Dawson! If you don't look out, you will be carried past!"

Eight days after Tappan Adney, Al Brown and others arrived, the Yukon River was frozen solid.

9

Starvation City

THE PREDICTION OF FOOD SHORTAGES had indeed come to pass. By autumn, the area's population had tripled to close to 5,000 souls. There was simply no way that the less-than-reliable riverboat system could provide the food needed for the sudden influx of hopeful and hungry prospectors. With winter on the doorstep, starvation loomed.

Dawson City, Yukon
Fall 1897

Charles Constantine was a frustrated, weary policeman. Since George Carmack's gold discovery, the newly promoted superintendent had been beset with one trial after another. For one, Bishop Bompas had attempted to persuade the overworked detachment to protect the local Native peoples from those who shot their dogs and sold them whisky. Constantine's common-sense advice to them had been simple: tie up your dogs and don't drink. His recommendations had not gone over well.

Bompas had moved the Han village away from Dawson City construction, only to have them move over to property set aside for the new police post, Fort Herchmer. Constantine suspected that the "arrogant Bishop" was behind their actions. Constantine wouldn't stand for that kind of coercion. The Native peoples were finally relocated five kilometres downstream.

Erecting Dawson City's new post had been made even more difficult by the fact that Joe Ladue's sawmill had not yet become operational. All logs had to be hand-hewn. Still, three buildings could be floated upriver from his headquarters at Fort Constantine.

Dawson City itself had evolved from an untidy hamlet of a few hundred souls to a centre of approximately 2,000 citizens—not all of them law-abiding. Constantine had concluded that no one—and no one's property— was safe. Thieves had even broken into the gold commissioner's office, located in the middle of the police compound. There were so many thieves and misfits, Constantine figured he could have filled cells at both Dawson and Cudahy. That is, if he had the men to round them all up and enough food to feed them.

Dawson was no longer the only Klondike settlement. One female entrepreneur, Belinda Mulrooney, had defied all the naysayers and built a two-storey log hotel about 20 kilometres from town, near where the Bonanza and Eldorado Creeks met. Others soon understood what she had instinctively known: exhausted miners wanted their goods and services close at hand. Soon other buildings lined Eldorado Creek, and by early fall the new town had taken the hotel's name for its own. The creation of Grand Forks stretched already thin police resources.

Word had come that the new commissioner of the Yukon was travelling north to take charge of all police. Members of a labour-saving administration team that included a supreme court judge, mining inspectors and a land agent were on their way. The Klondike's transformation from rude gold camps to a civilized northern centre appeared to be at hand. Fortunately, the commissioner was a well-known former policeman with a sterling track record: Major James M. Walsh. Unfortunately, winter arrived before he and the others did. Walsh got the farthest, but he was forced to stop upriver at the Big Salmon police post. A gold commissioner had been at work there since the summer, but the inept man inspired so many complaints that he was more a hindrance than a help. So, as before, most of the work overseeing justice and administration in a rapidly growing area continued to fall to Charles Constantine and a mere 40 policemen.

To be sure, there were more police in the North now.

In the spring of 1898, the Dawson waterfront was still essentially a tent town—but not for long.

A detachment in the US town of Skagway, where Mounties wore civilian clothes to maintain a low profile, kept supply and information lines open. Posts had been built at Lake Bennett, Tagish Lake and White Horse Rapids. Still, men in these far-flung outposts were of little strategic value to Constantine in Dawson.

In October, Constantine had been delighted to welcome an additional inspector and 20 extra reinforcements—until he realized they had arrived without supplies to feed themselves or anyone else. By this time, fear of food shortages was almost palpable. Both Jack McQuesten and John J. Healy had built stores and large iron-fronted warehouses, but Constantine was not misled. He wrote to the superiors with the warning that "many deaths . . . will result from starvation and privation during the coming winter."

The first hint of trouble came in September. A steamer had not been seen for weeks. The Alaska Commercial Company's assistant superintendent, Captain J.E. Hansen, headed downriver to discover why. He discovered three boats marooned by low water levels about 560 kilometres north of Dawson, near Fort Yukon. When he arrived back at Dawson City, 4,000 people were waiting. When the hoped-for steamer turned out to be Hansen's lone canoe, the tense crowd grew ugly. Captain Hansen announced fearfully, "Men of Dawson! There will be no riverboats here until spring." He advised people to leave town immediately for Fort Yukon, where, he said, the provisions they needed were available. "There is no time to lose," he exhorted. Women and men screamed their outrage. Panic spread.

Men flooded into Dawson City from Bonanza and Eldorado Creeks. Supplies had to be brought upriver from Fort Yukon. Partners drew straws to see who would be forced to go and who would stay. Fifty boatloads of men left for Fort Yukon. When they arrived at their destination, shocked prospectors learned that gold had very little influence with store clerks, who doled out meagre quantities and carefully locked the door as each customer entered.

Meanwhile, in Dawson City, John J. Healy pleaded for calm and patience, convinced that boats would arrive. To fill the void, he began distributing two-week allotments of supplies from a rough platform on muddy Front Street.

Dawson City, Yukon
Winter 1897

As yellow leaves fell from the poplars, there was talk of storming the Alaska Commercial Company and NAT&T warehouses. The commanding officer, Sergeant-Major Davis, offered to place policemen at the warehouses to safeguard supplies. Hansen accepted the offer. Healy refused. A show of force might only incite violence, and violence, Healy knew, could mean the loss of NAT&T property and inventory. Moreover, Healy thought, the issue wasn't food shortages, it was food *distribution*. Healy perceived that this so-called crisis contained a hidden opportunity.

Healy's warehouse contained plenty of everything except flour—the miners' staple. However, Captain Hansen's company warehouse held plenty of flour. Here was a chance to rectify that problem. Healy approached his competitor, eager to make a deal. "If you will let me have flour, I can let you have bacon, sugar, everything else," he suggested to Hansen. "I have one-year orders from $500 to $10,000. You have the same. My proposition is this: fill every man's order as nearly as possible, but," and Healy felt this was critical, "cut them down from twelve to eight months." By then, river traffic would resume. Some might go hungry in the meantime, Healy admitted, but no one would starve.

Hansen thought it over briefly and then said, "I must fill my orders." Healy left in disgust.

When the *Portus B. Weare* stopped at Circle City, desperate men who had watched boat after boat float by the riverside hamlet met the captain and crew with drawn guns. This boat wouldn't move again, they had decided, until they had what they needed. Money was no problem. They had the money. What they wanted were supplies. They off-loaded 27 tons of supplies at gunpoint and then cheerfully paid for everything they had taken off. The *Bella* was similarly hijacked. She left for Dawson City 23 tons lighter.

A few days later, the sound of a boat whistle floated across Dawson City. Throngs came running to see the tall structure of the *Portus B. Weare* silhouetted against the slate-coloured sky. John J. Healy shouldered his way aboard. Breathlessly, he asked the NAT&T president and Portus Weare's brother-in-law, Eli Gage, to confirm his cargo. Gage reported the ship was carrying every bottle of whisky and case of hardware they could float across the flats. Healy's mouth flew open in disbelief. He had given orders to load only food and clothing! Enraged, Healy lunged forward and wrapped his hands around Gage's throat until one of the ship's men pried them loose.

At the end of the month, when the half-empty *Bella* reached Dawson City, Inspector Charles Constantine was

The Scales on the Chilkoot Trail, winter 1897–98.

aboard. Reports reaching him at Fortymile had convinced him that a firm hand was needed upriver. After quick consultations, Constantine posted a bluntly worded public notice, urging those without food for the winter to get out. The notice made the alternative frighteningly clear: "Starvation stares everyone in the face who is hoping and waiting for outside relief."

John J. Healy was nonplussed; leaving the warmth and safety of Dawson City as winter approached was crazy. When miners organized a meeting to discuss the evacuation, Healy refused to attend, sending down a representative instead. There was plenty of food, he argued. The more he pleaded for reason, the more the animosity toward him grew.

Three hundred people departed. Many paid $50 for passage out on the *Portus B. Weare.* The next day, 160 left on the *Bella* as guests of the NWMP, paying nothing at all for the privilege and receiving a week's food in the bargain. Ironically, both sailings proved to be profitable opportunities for Healy. Business was business, and Healy didn't care whether fares and food were paid by prospectors or the police. A small armada of private boats and scows quickly followed down the river. It seemed that Superintendent Constantine had solved Dawson City's problem. On the river, problems for evacuees were just beginning.

After ice hemmed the ships in at Circle City, a sudden chinook cleared the water. Dozens of boats set out for the 136-kilometre trip to Fort Yukon. Twelve hours later, the temperature plunged, and the ice was back. Trapped, splintered boats littered the frozen river. Several hundred destitute, starving men staggered along for three days in sub-zero temperatures. At Fort Yukon, contrary to assurances, the hungry wanderers were told there simply wasn't enough food to go around. Anarchy reigned; the warehouse was ransacked and thousands in gold was stolen.

Downriver at Circle City, some people had already decided to return to Dawson City. In temperatures as low as −50°C, thousands stumbled toward the Klondike. Bent and

Policemen, such as those clustered around the Union Jack, kept Soapy Smith's gang at bay, to the relief of thousands of stampeders.

ravaged, they climbed over and around gigantic hummocks of ice, littering the frozen surface with abandoned sleds and packs and shivering through the nights inside small tents.

Day after day, hundreds of misguided, suffering wretches stumbled on, desperately cold and hungry, toward Dawson City—the very place they had been persuaded to leave months before. Many feared they would die in the howling darkness long before seeing the warm glow of the lamps and lanterns of civilization.

All winter long, in twos and threes, the hundreds of destitute travellers shuffled off the snow-covered ice of the frozen waterway, climbed the riverbank and plodded numbly down Front Street, begging for food and shelter. The worst, they surely felt, was now over; they were safe, at last. It was not to be.

For some of these tortured individuals, and many others who'd never even left starvation city that cruel winter of 1897, the pain and suffering were far from over. Prostrate prospectors lay moaning feverishly on their cabin bunks and inside the city's overcrowded log hospital, unconscious from exposure. For some, death came as a relief, taking some of them with pneumonia, coughing and gasping their last breath. Others, groaning in pain, pried loosened teeth out of their mouths before they succumbed to the ravages of scurvy.

CHAPTER

10

The Law
and the Lawless

REPORTS OF THE ENORMOUS GOLD finds a few months before and the instantaneous frenzy that resulted stunned people in Ottawa, as it did in every other sizable city in Canada and the United States. However, in Canada's capital the sudden activity was not limited merely to the brave few who left to strike it rich.

Federal government officials who had speculated leisurely on the effect of a potential gold rush were now galvanized into action. Their particular mission was safeguarding the sovereignty of this remote corner of the nation from the sudden onslaught of foreigners, which—according to NWMP superintendent Charles Constantine—included the most lawless element of American society. The government dispatched more North-West Mounted Police. In the fall of 1897, it appointed a former Mountie hero, James Morrow Walsh, first commissioner of the Yukon and NWMP superintendent, to replace worn-out Charles Constantine. Ottawa sent the legendary Walsh north with a "court party" that included a judge, a crown prosecutor and other court officials.

A Tarnished Legend

Upon hearing that Walsh was on his way to maintain order in the Klondike, many concluded he was just the man for the job. They recalled the Mountie's dashing exploits on the Canadian prairies, over a quarter of a century before, when Walsh's no-nonsense approach to the arrival of the Sioux had earned him the nickname "Sitting Bull's Boss." Alas, now 54 years old, Walsh was no longer the man they remembered.

Even as the party laboured along the trail to Dawson City, some were having doubts. "He used to be tough," Assistant Commissioner J.H. McIllree wrote, "but he is not as young as he used to be." Nevertheless, Major J.M. Walsh was now the "Klondike's Boss." Ironically, the very first person to rue that fact was none other than a fellow NWMP officer.

On the White Pass, the raw reality of the gold rush appears to have made an impact on "spit-and-polish" Walsh. "The inhumanity which this trail has been witness to, the heartbreak and suffering which so many have undergone, cannot be imagined. They certainly cannot be described," he reported.

Conditions in the North appeared a lot more primitive than at Fort Walsh, where, pennants flying, the white-gloved major had once led scarlet-clad horsemen against the Assiniboines. To Walsh, appearances reflected discipline and professionalism, and the appearance of Yukon policemen was, he wrote, "very disappointing, their costumes as veriegated [*sic*] of those of the packers on the Chilkoot Pass." He accused the men under his charge of being "the laughing stock of people all along the trail."

Walsh seemed to ignore the effect harsh realities had on his underpaid, overworked Yukon policemen. Not surprisingly, the first contingent he met on the other side of the mountains did not meet his parade-square expectations. Inspector D'Arcy Strickland and his handful of sweaty, dirty tent dwellers had been working for weeks from dawn until dusk to hack the new Tagish River post out of the bush. The commissioner bluntly reprimanded Strickland for his "scruffy looking men."

A graduate of the Royal Military College at Kingston and fully cognizant of the commissioner's sweeping powers, pragmatic Strickland likely replied, "Yes, sir!" and was not unduly upset over Walsh's pompous arrogance. Strickland, who built Fort Constantine, the Yukon's first NWMP post, understood conditions far better than Walsh. The commissioner would learn, and the lessons would come hard.

Others were not so sanguine. At the newly constructed Dawson City post, where a growing number of prisoners awaited the judge who would preside over their trials, policemen might well have asked, "Where the dickens is Walsh, anyway?" The answer: snowed in at the tiny police post at Big Salmon, hundreds of kilometres south, where the commissioner scribbled diatribes about police inefficiencies. Walsh wouldn't arrive at Dawson until the warmth of spring broke the ice on the Yukon River. Seemingly helpless in the face of the elements, Walsh appeared to personify his own police "inefficiencies."

Nevertheless, one member of Walsh's party, prosecutor Frederick Wade, did manage to reach Dawson that winter and later claimed the commissioner could have done the same. The reason for Walsh's delay, Wade told the secretary of the interior, was "sheer fear." Walsh staff member J.D. McGregor put it even more bluntly. Walsh, he said, was "utterly devoid of courage."

What Wade and McGregor failed to note was that Walsh had started out for Dawson by dogsled in early February. However, he learned that in response to the threat of starvation, a US relief expedition accompanied by American troopers was poised to enter the Yukon. Walsh doubled back toward the White and Chilkoot passes to meet this intrusion into Canadian soil. In a move that saved countless lives, the commissioner issued an order that no one would be allowed to enter Canadian territory without at least one year's provisions.

Meanwhile, in Dawson, simmering discontent over alleged corruption in the gold commissioner's office boiled over into outright rage. The frustration began with unpopular Canadian mining regulation royalties that rose to 20 percent when weekly earnings exceeded $500. When 3,000 angry miners met to protest, the gathering ended in fist fights. Chaos ruled as the government's inadequate staff attempted to manage a deluge of mining claims. Gold Commissioner Thomas Fawcett became the focus of the anger. The *Klondike Nugget* was quick to lead the crusade, and publisher Gene Allan recalled that the newspaper "sold like hotcakes."

It wasn't long before suspicions of corruption surfaced. The date previously announced by Fawcett for the opening of Dominion Creek claims was suddenly put ahead. However, a select few close to the gold commissioner's office got advance word of the change and were already on the creek, poised to stake their claims. More outrage!

"The administration . . . on Dominion creek has been a mess from start to finish and I am sick and tired of the whole business," Walsh confessed to the *Nugget*.

GOODBYE FAWCETT shouted the headline at the gold commissioner's hasty departure for the outside. Soon the taint of corruption spread to the commissioner himself. Walsh's former cook testified that in order to get his job, he was forced to sign over two-thirds of any mining claim he staked to either James Walsh or his brother, Philip. Even though results of a formal investigation headed by William

Ogilvie were inconclusive, Walsh's career in the Yukon was over, and he left under a shadow of suspicion.

Months before, government officials in Ottawa realized that sending Walsh was not enough. The Yukon needed a single, stalwart individual with proven leadership and steadfastness in the face of crisis. Their thoughts turned to one particular Mountie, Superintendent Samuel Steele.

Fort Macleod, Alberta
January 1898

Far from the chaos of the Klondike, Samuel Steele celebrated the new year in Alberta with his friends, wife and three small children. Amidst the toasts for peace, health and happiness, Steele was resigned to the probability that 1898 was likely to offer the same humdrum set of duties as the year before. Compared to the days of whisky traders, the battles of the North-West Rebellion, Native treaties, murder investigations and a CPR construction-camp riot, the day-to-day affairs of Fort Macleod seemed an anticlimactic finale to a distinguished police career that had begun with the NWMP's famous March West. Steele might have stoically concluded his fate was simply "just desserts" for his intolerance of incompetence at any level, including at the very top.

Steele had been passed over for promotion as commissioner, a position granted to William Herchmer, a civil servant with scant military experience. Before long, Steele ran afoul of the pompous and dictatorial Herchmer. The commissioner's campaign of personal harassment against Steele escalated to spiteful accusations of misconduct. Steele stood his ground and Herchmer backed down. As a consequence, Steele lost another chance at promotion. The position of assistant commissioner went instead to Herchmer's own brother. Steele soldiered on.

The attitude of the new federal government was cause for concern. While in opposition, the Liberals had taken the Conservatives to task about the NWMP, claiming it was an expensive exercise in patronage. The prevailing opinion was that since the West had been tamed and the railroad finished, the days of the NWMP were over. The Liberals were now in power. The future did not look bright for Steele or the police force itself.

On January 29, a telegram from Herchmer arrived at Fort Macleod. Superintendent Samuel Steele was stunned. He had been ordered to the Yukon. It seemed that the force's day was not yet over. Nor, for that matter, was Samuel Steele's career.

Chilkoot and White Passes
Winter 1897–98

Through snow, wind and cold, hopefuls continued to labour up the Chilkoot and White passes, mounting the 1,200 icy stairs of the Chilkoot, or tramping up past grotesque heaps of grinning animal corpses on the White Pass. By late

February, men hardy enough to make the summit found the NWMP waiting.

Two wooden crates that had been secretly stashed at the rear of the police post at Lake Bennett two months before were dug out of the snow and pried open. Officers lifted out the very latest in turn-of-the-century weapons technology: two British-made Maxim machine guns. Constantine had pleaded for the guns as early as 1896, fearful his small force would be overrun by "toughs, gamblers, fast women, and criminals."

As if to underscore Constantine's concerns, the force's small Skagway office found itself in the center of a gunfight. The sound of gunfire was not all that unusual in Skagway that winter. On this occasion, however, all officers hit the floor when bullets began to ricochet through the office itself. It wouldn't be the last time, nor were they the only ones to experience gunplay at close hand.

"The shack I slept in had a bullet through it over my head," reported one experienced traveller who had seen many frontier towns. Lawless Skagway made a lasting impression. "For the six nights I slept in Skagway there was a shooting on the streets every night. At least one man was killed that I knew of and probably others."

At last Ottawa responded, sending "equalizing" armament designed to put a tiny group of police on the same footing as hundreds of outsiders who might mount an armed uprising. Ottawa also instructed officers to be posted at the summits. It wasn't the fear of lawlessness that prompted the government to act; it was the potential problem of huge numbers of foreigners posing a threat to the nation's sovereignty. At the time, journalist Tappan Adney reported as many as 90 percent of the miners in the Canadian territory of the Yukon were from anywhere but Canada.

Upon reaching the Chilkoot's Scales or the White Pass' Summit Lake, no individual could proceed farther without proving they had a year's food and supplies. James Walsh's edict, soon to be backed up by Sam Steele, meant that every person on the trail was required to push, pull or carry approximately one ton of sacks and boxes filled with an astounding array of goods, including 23 kilograms of evaporated onions, 181 kilograms of flour, stoves, tents and even underwear. Millions of tons of goods were moved over the Chilkoot and White passes before the fall of 1898.

Travellers hiked up and down the trails repeatedly to bring up their supplies. Later, for those with the means to pay, an aerial tramway made loads lighter and the ordeal shorter. At the summit, supplies were often obliterated by overnight snowfalls. The roof and the top third of the flagpoles were frequently the only visible reminders of the police post buried beneath.

To facilitate the orderly travel of thousands of stampeders, American army officers in Skagway and the Canadian police on the other side of the mountains came to the amicable—

and practical—agreement to set the international boundary at the summit of the passes. While the final boundary would be debated until 1903, for the time being, the fluttering Union Jack signified Canadian sovereignty over the territory and facilitated the collection of customs duties. By June 1899, the total duties collected exceeded $174,000. Earlier, the onerous collections were made worse by numbing winter weather. Said one frustrated officer, "No one with a spark of humanity would keep people waiting in those dreadful places, with the danger of perishing from cold, while their goods, exposed to the inclement weather and blowing snow, spoiled before their eyes."

Exercising common sense and good judgment, the policemen earned the respect of thousands of travellers and the admiration of their new commander, Superintendent Sam Steele. Steele reported, "The officers in charge of the summits displayed great ability, using great firmness and tact." Men in the ranks were recognized too, for "the greatest fortitude and endurance amidst the terrific snowstorms."

One storm was so fierce that the policemen retreated temporarily from the summit on Chilkoot and set up camp below at Crater Lake. During the night, the water level rose, tents were flooded and everything was soaked.

Keeping customs men warm at the summit meant cutting cords of wood. Firewood details did not dare chop down the nearby trees. Those were American trees, and nobody wanted to risk an international incident. So, frost-bitten men trudged kilometres up and down slopes, cutting and hauling wood from the Canadian side.

Improvisation and snap decisions were the order of the day. Steele confiscated supplies and a revolver from a suspected American thief and sent him under guard down the trail. One observer was stunned by Steele's audacity. "How can you do that?" he asked incredulously.

"I *can't* do it," Steele admitted, "but I'm not going to have any of those thugs robbing and murdering on this side like they are doing down in Skagway." He smiled slyly. "I've been waiting for just such a one over here to make an example of him."

Skagway was, Steele believed, "about the roughest place in the world. Robbery and murder were daily occurrences; many people came there with money, and next morning had not enough to get a meal, having been robbed or cheated out of their last cent . . . occasionally some poor fellow was found lying lifeless on his sled . . . powder marks on his back and his pockets inside out."

In Skagway, the man ultimately responsible for much of the murder and mayhem was Soapy Smith. That past August, Smith had decided chaotic, transient Skagway was his kind of town, and he quickly made it his own. The ranks of Soapy's gang swelled to hundreds of vest-and-derby con artists, cloth-capped thugs, cold-hearted pimps and gaudy prostitutes. His arm's-length associates included the deputy marshal, justice of the peace and a newspaper editor. Disguised as hopeful prospectors bent under heavy-looking packs, Soapy Smith's armed bandits "struck it rich" at the expense and sometimes the lives of others. A keen eye, steely nerve and quick thinking were often the best ways to foil the gang.

Skagway, Alaska
May 1898

William "Cap" Olive was the manager of the newly established Bennett Lake and Klondyke Navigation Company. He travelled to Skagway to cash a Bank of Montreal draft to pay freight handlers and the men who were building the company's three steamboats at Lake Bennett. Once he entered town, Olive knew he was being followed. At the First Bank of Skagway, the overly loud replies of the suspicious-looking teller and manager seemed designed to be overheard by men loitering nearby. Olive wanted notes, not gold, for his journey. The manager asked Olive to return later that afternoon. When Olive left the bank to visit his agent, two of the men from the bank followed him. His agent spotted them through his office window.

"Soapy's lookout," he told Olive. "For God's sake be careful. They're known as the Weasel and the Panther. Don't take any risks when packing the money to Lake Bennett. They'll get you as sure as the Lord made little apples."

Olive realized it made little difference whether he was in town or on the trail. "If I don't pack it, they'll still get me." He had promised his workers their money without delay, so he decided to take the risk. Before leaving, Olive called in the deputy marshal and related his story. Olive suggested that bank personnel were in Soapy Smith's pocket. The marshal was belligerent rather than sympathetic.

"It's a damned good thing I know you, Cap, or you couldn't talk to me like that about the bank," the marshal snorted. "You're imagining things."

"Oh no, I'm not," Olive replied, realizing at once which side of the law the marshal was on. "Watch your step, marshal," he warned. Minutes later, Olive and his agent saw the marshal brazenly talking with one of the gang members.

All that day, Olive could feel eyes watching him. That night, inside the hotel restaurant, he and his agent performed a loud charade for the benefit of gang members sitting nearby. "Here's off for a sleep, Jack," Olive announced, getting up from the table. "Wake me at 2 a.m. and we'll be off on the White Pass Trail. Goodnight!" Instead, within the hour, Olive was being rowed toward Dyea, headed for the Chilkoot Pass.

Once at Dyea, Olive knew that he was being followed again. Hugging his bag of bank notes and $1,000 in gold double eagles, Olive dashed for the Taiya River. The log bridge was out; he would have to cross on foot. Up to his knees in rushing water, Olive heard a gunshot behind him. He pulled out his revolver, whirled and fired in the direction of the muzzle flash. He heard a moan in return. As promised, Olive's workers received their pay.

Later that night in Skagway, a chance meeting with a

NWMP officer resulted in the rescue of Olive's agent. Minutes earlier, he had been apprehended at gunpoint by Soapy Smith's thugs, eager to beat him senseless for losing Olive.

* * *

In June, customs collections amounting to $130,000 in gold and notes were sitting at the NWMP Lake Bennett detachment. Sam Steele, who had made the busy detachment his headquarters, asked Inspector Zachary Wood to make the trip down the trail, through Dyea to Skagway, to see the money safely on to Victoria.

The policemen knew they needed a strategy to elude Soapy Smith's mob. The men arranged to spread the word that Inspector Wood was being transferred to the Canadian prairies. The story made sense as Wood had recently closed up his Skagway office. Wood and two officers hit the trail.

The men now carried a huge sum—$224,000—stuffed into two kit bags. At Dyea, they booked a hotel room and guarded the cargo in shifts. In spite of the cover story, word came that the officers were being watched. It was as though Soapy Smith could smell the money.

Wood formulated a strategy. A telephone call confirmed that the *Tartar* had arrived in Skagway. They had to board it at that location. So, at the beach, Wood and his men uncovered a hidden rowboat, and within a few minutes they were safely aboard a rented tugboat, proceeding down the inlet to Skagway. But the ordeal wasn't over yet.

A rowboat full of Soapy Smith's men bore down on the tug. The officers shouldered their carbines. Wood's voice echoed out over the water, warning Soapy Smith's men to keep their distance. The pursuers veered out of bullet range but continued to follow. Aboard the tug, the policemen could see the *Tartar* tied up at the wharf. The crew was at the hurricane rail, armed with rifles. Pushing and shoving, Soapy Smith's men closed in on the policemen as they stepped up onto the wharf. Bags in hand, Wood found himself face to face with Soapy Smith.

Suddenly, from the *Tartar*'s deck came a hoarse command, and a contingent of uniformed navy reservists moved quickly onto the wharf, rifles at the ready. Outgunned, Soapy Smith realized that this was one gamble he could not win. Bowing slightly, he smiled at Wood. "Why don't you stay awhile, inspector?"

Inspector Wood graciously declined Smith's offer of "hospitality," and his group carried the precious cargo up *Tartar*'s gangway.

Lake Bennett to Dawson City, Yukon
Spring–Summer 1898

A month before the Mounties' American showdown on the Skagway wharf, Salome and Tom Lippy were living with their new baby in a tent at Lake Bennett. It had been a wonderful time away on the outside, but now, like the Berrys, the Lippys were back in the North. The Berrys had sledded across the frozen lake a few weeks before. The Lippys and thousands of others were anxiously waiting for spring breakup to float down the river. All day long the sound of saws and hammers filled the air as hundreds of boats took shape on the broad, curving shoreline.

The homemade creations hammered together in the huge Lake Bennett tent town often reflected the builders' haste (the vessels soon sprang leaks and quickly foundered), and sometimes their sense of creativity. Stampeder Mont Hawthorne knew about boats from his days at the Columbia River canneries and built a scow sturdy enough to hold himself, two others and their tons of supplies.

By the end of May, the ice was gone. Superintendent Sam Steele walked up the slope behind his office and watched as an astounding 800-boat armada sailed off into the distance. Mont Hawthorne saw strange vessels the likes of which he had never encountered. "One was built square by a fellow who was going down alone," Hawthorne remembered. "He had two sets of oarlocks, and he had put one in each side. When we passed him, he didn't seem to be making much headway frontwards. But the funniest boat I ever did see was one four fellows built up near us on Bennett. They had come up the Mississippi from New Orleans on an old side-wheeler, so they built their boat with two small side-wheels instead of using oars. Each wheel was on a crank, and they took turns playing engine."

Among the first to push off were the Lippys, along with

two male companions. Their pleasant sail ended about 70 kilometres below Lake Bennett. Barely into Miles Canyon, the scow smashed up onto a half-hidden rock. It was stuck fast. Crouching down in the boat, Salome clutched her baby tightly. A quick inspection revealed that the scow was not damaged. As the water surged around them, ropes were thrown around nearby boulders, and heaving with all their might, the men refloated the scow.

Moments later, the heavily loaded boat turned alarmingly and then began to spiral out of control, sucked into a deadly whirlpool. The bow dipped dangerously as the frantic men strained at the rudder and their oars. Gradually, the scow floated free. There was no time to think as the craft raced through the foam of White Horse Rapids.

With a splintering of wood, a jarring concussion tossed the passengers about. Horrified, the Lippys and their friends gazed at the damaged bow. With an agonizing scream, one of the men threw himself overboard and was quickly swept away, out of sight. Grabbing an oar and the rudder, the remaining two men eased the damaged boat off the rock and carefully steered through the white water to a sandbar. Shaking and breathless, the group staggered ashore without so much as getting their feet wet. Others fared far worse.

The Lippys' terrified companion was one of 10 men who disappeared in the foaming waters. Sam Steele estimated that about 150 boats and their cargoes had been smashed to pieces on the treacherous river. He decided the time had come to stop the carnage. At the Whitehorse Rapids detachment, Steele asked his men to assemble as many of the travellers together as they could. Steele knew that most of his audience would be American. This was a good opportunity to let them know who was in charge and what they could expect. There would be no apologies.

"There are many of your countrymen who have said that the Mounted Police make the laws as they go along," Steele shouted out to the crowd, "and I am going to do so now for your own good."

There were to be no more women and children in boats. Instead, they could walk the short trek around the rapids. Fines of $100 awaited anyone who flaunted the regulations. Corporal Edward Dixon, who happened to be an experienced river pilot, was ordered to assess the competency of all those at oars and rudders and appoint substitutes if necessary, Steele told the crowd. Standing next to the skippers inside steamers' wheelhouses, the corporal steered the vessels through the rapids. He never lost a boat.

Before the year was out, Sam Steele was comfortably housed in a cabin at Dawson City. Here, NWMP foot patrols were conducted 24 hours a day, and a good number of the police force's "guests" housed in jail cells were employed in the Royal Fuel Factory, swinging axes and pushing saws to replenish the huge woodpile.

"Robert Russell got 18 months on the woodpile and really deserved more," the *Klondike Nugget* told its readers. The man was "broke and sick," but was nursed back to health by the police surgeon and then employed in the officers' mess. How did he repay the Mounties? "By stealing everything in sight," the newspaper reported. "The woodpile at temperatures of 50 below may work reformation."

During the summer, woodcutting became clean-up duty, 10 hours a day, 6 days a week. As Dawson City's lawless discovered, the wages of sin were hard-earned, indeed.

But what was to be done about the town's twin vices, gambling and prostitution? These were activities that few civilized cities of the time countenanced—at least, officially. Yet, this was Dawson City, where life was harsh, toil was even harder and refined women still a tiny minority. Steele knew full well that gambling and prostitution were "seen in the eyes of the majority of the community a necessary evil," as he reminded his superiors at NWMP headquarters in Regina, Saskatchewan. The NWMP allowed the dance halls, gambling dens and saloons to operate without much restriction. However, Steele warned saloonkeepers that if he heard about cheating, he'd close them up. About the only other time these doors were closed was on Sunday, when virtually all commerce shut down.

Prostitution was a particularly thorny issue. At first, the prostitutes operated out of tents throughout the town. Steele had them congregate in one location, just off Front Street, behind dance-hall row. The growth of the town meant that the cribs along this "Paradise Alley," were situated on increasingly valuable real estate. At the urging of property-hungry merchants, Steele ordered the ladies moved to a more remote location, a bit of swampland which became known as "Hell's Half Acre," where they continued to ply their trade.

The Mounties' worldwide reputation for dogged determination and understated courage in the face of danger was firmly established during the Klondike Gold Rush. One of the many small dramas played out in the Yukon between lawman and criminal was witnessed by dog puncher Arthur Walden, who was working a claim on Last Chance Creek, about 30 kilometres from Dawson.

No one knows what sparked the argument between two prospecting partners, but before the day was done, one had shot and killed the other. Many heard the shot ring out and watched the killer dash into his cabin and bar the door. His threat was loud and clear: he would kill any man who came near. All work stopped, and while angry, armed miners watched the cabin from behind trees and brush, a volunteer raced away to Dawson to bring back the police.

When a solitary constable finally arrived, Walden and others warned him that it was "certain death" to get within shooting range of the man. The stout log cabin topped with a foot of dirt on the roof was, Walden figured, "a first-class fort. All he needed to do to make a loop hole was to poke some moss out from between the logs."

Tired after his long trek, the Mountie sat down and had a smoke. Then, while the tense crowd looked on, the constable stubbed out his cigarette and sighed, "Well, I'll guess I'll have to make a try at it." The Mountie simply strode up to the cabin, pounded on the door and ordered, "Open, in the Queen's name!"

"Every gun of the watchers was turned on that door," Walden recalled, "law or no law, I think every man intended to shoot the occupant if the policeman were killed."

Instead of blazing away, the killer threw open the door and stepped out with hands extended, ready to be hand-cuffed. The Mountie hadn't even drawn his revolver. The stunned prospectors couldn't believe it. "The whole thing was done with about as much spectacular display as if the policeman had been looking for a piece of string to tie up a dog with," Walden marvelled.

Disbelief deepened when the prisoner confessed to the constable that he simply waited, "till you got near enough so I couldn't miss you. I had every intention of killing you . . . but I found it impossible to shoot down in cold blood a man who was braver than I." He then turned to the slack-jawed men gathered around him and told the Mountie in no uncertain terms he "would have liked to get in a few shots at these blankety-blank cowards who had me surrounded, only they were hardly worth it."

Watching the constable start down the trail with his prisoner, Walden concluded, "This was the method of the North-West Mounted Police: one man for a man. But they had the majesty of English law back of them."

The Beginning of the End

THE KLONDIKE'S TENS OF THOUSANDS of stampeders left behind jobs, homes and families to chase their gold-rush dream of instant wealth. For all but a select few, it was a dream that would never come true. All the best gold-bearing dirt had already been claimed long before any of the dreamers even began their journey north. Unfortunately, none of the would-be prospectors learned this terrible truth until they had travelled more than 1,000 kilometres to Dawson City. By the time the first of them arrived in the early fall of 1897, there was scarcely a decent gold-bearing creek left unclaimed in which to dip a pan.

Saying Goodbye

Dawson City and Grand Forks thronged with thousands of men. Feet dangling in the dirt, they sat in rows on the edge of raised wooden sidewalks with nothing to do and little to live on. Some helped others find gold as paid labourers, working the claims of millionaires. Men outnumbered jobs and crew wages plummeted. By the fall of 1898, the exodus back to civilization had already begun. Some of the Klondike's most well-known figures joined the thousands who had already departed. One was Superintendent Charles Constantine.

Constantine's four-year term in the Yukon had been a period of astounding change and overwhelming demands. His first small police contingent of 20 officers had grown to a force of 264 manning a network of 31 posts. As Constantine and his family floated down the Yukon River, he paused to make a brief, but undeniably heartfelt entry in his diary: "Left Dawson per str. C.H. Hamilton. Thank God for the release."

Ironically, one man who stayed while so many others left was someone who'd regarded his stay in the Yukon as merely a temporary posting. With the unexpected removal of James Walsh, William Ogilvie, the one-time government surveyor, was thrust into the position of commissioner. Ogilvie, the incorruptible civil servant, later returned to eastern Canada and became an author, writing one of the gold-rush era's most valuable narratives, *Early Days in the Yukon*. He died a pauper.

By the summer of 1898, reporter Tappan Adney had filed his last Klondike story. Adney recorded his adventures in *The Klondike Stampede*, a book still read avidly today. By the time it was published, however, the Klondike Gold Rush was "old news." Initial book sales were not large. Adney's lesser-known legacy remains his most important. For years he crafted dozens of meticulous scale models of Aboriginal bark canoes. These models and his painstaking notes about their design and construction tell us almost everything we know about this now-vanished cultural art form. In 1950, Adney died in near poverty in New Brunswick.

In September 1899, Superintendent Samuel Steele was suddenly dismissed from the Klondike. Steele had run afoul of his superiors again. The *Klondike Nugget* pulled no editorial punches, claiming Steele's removal was the work of Minister of the Interior Clifford Sifton and "the Sifton gang of political pirates."

Steele was revered by the people he protected. Thousands of grateful Klondikers saw him off at the wharf. Alex McDonald was asked to present Steele with a gift of gold nuggets. "Here, Sam," the taciturn millionaire mumbled, "Here y'are. Poke for you. Goodbye." The *Klondike Nugget* was much more fulsome, calling Steele "a man of spotless reputation," and "the most highly respected man in the Yukon today."

Sam Steele's career was far from over. The Boer War brought him the opportunity to organize a special mounted

unit, Lord Strathcona's Horse, which added more lustre to
the reputation of both the man and the country he'd served
so well. Steele died in England. His Winnipeg funeral—the
largest that western Canada had ever seen— put an abrupt
halt to the 1919 Winnipeg General Strike. Police and strik-
ers alike paused to honour the man who'd personified law
and order in Canada's far-flung frontiers.

The Future of the Fortunate Few

Some millionaires outlived their riches. Thomas and Salome
Lippy sold their enormously rich Klondike claim in 1903,
returned to Seattle, built a grand home and began to invest
their fortune. The couple's streak finally ran out when the
Great Depression rendered their investments worthless. The
couple died bankrupt and impoverished.

After touring Paris and Rome and enjoying an audience
with the Pope, Big Alex McDonald continued to use successful claims as collateral against other purchases, most of
which turned out to be worthless. Eventually, his house of
cards collapsed. Years later, the body of the impoverished
"King of the Klondike" was discovered lying in the place
where his heart had stopped. McDonald had been sawing
firewood outside his little cabin in the Stewart River region.

Others, including John J. Healy, Jack McQuesten and
Clarence and Ethel Berry, lived out their long lives in lux-
ury. The Berrys struck it rich three times. Leaving the Yukon
wealthier than they could have imagined, they found more

riches in Fairbanks, Alaska. Later they turned their yellow gold to "black gold" and founded California's Berry Oil Company.

Others, however, did not live out the gold rush. Joe Ladue finally married his long-time sweetheart, Anna Mason. She was still waiting for him in the spring of 1897 when he travelled back to the eastern United States. Anna's wealthy family welcomed millionaire Ladue as one of their own. Wedded bliss was short-lived, however. Twenty years of wilderness hardship had taken their toll. Ladue died of tuberculosis less than a year later.

Jefferson "Soapy" Smith met his death at the barrel of a gun. In response to yet another theft, the leader of Skagway's vigilante committee, Frank Reid, went gunning for Smith. "There'll be trouble unless the gold is returned," a *Skagway News* reporter told Smith.

"By God, trouble is what I'm looking for," Smith shouted. He found more than enough of it on one of William Moore's wharves. Armed with a rifle, a revolver and a derringer, Soapy Smith stomped down to disrupt a vigilantes' meeting. Barring his way was Frank Reid.

"You can't go down there, Smith," Reid calmly told him.

"Damn you, Reid, you're at the bottom of all my troubles," Smith snarled, stepping forward. Now the two men were within arm's length of each other. "I should have got rid of you three months ago."

Soapy Smith levelled his Winchester at Reid. Frank

jerked the barrel downward and drew his own revolver. The hammer clicked on a dead cartridge as Soapy Smith's rifle sent a bullet into Reid's groin. Through his pain, Reid squeezed the trigger again. Smith stumbled back from a fatal shot to the heart, discharging his rifle again, hitting Reid in the leg. Both men staggered to the planking. Reid fired again, hitting Soapy Smith's left leg.

"I'm badly hurt boys, but I got him first," Reid mumbled to the gathering crowd. Soapy Smith was dead in minutes. It took Reid days to die. His funeral was the largest in Skagway's history.

By October 1889, Robert Henderson, arguably the Klondike's biggest loser, had left the Yukon. His slurs against Skookum Jim and Tagish Charlie had been, in part, the reason why Carmack never told him first about the strike on Rabbit Creek. Henderson's lengthy isolation on Gold Bottom Creek cost him dearly. In the weeks following his discoveries there, the mining laws had changed. When he reached Fortymile, Henderson was stunned to learn he could register just one of his three claims, one discovered less than 60 days before. The time factor alone made his Gold Bottom claims worthless.

"These discoveries rightly belong to me," he railed, "and I will contest them as a Canadian as long as I live!"

Henderson was anxious to leave the Yukon to rejoin his family in Colorado. Bad luck dogged him every step of the way. The bitter prospector was iced in at Circle City, where he became so ill that he sold one of his former claims to pay medical bills. When he arrived in Seattle, he was virtually penniless. A shipboard thief had stolen what little gold Henderson had left.

By pure chance, Henderson bumped into Tappan Adney in Seattle. He told the reporter his anguished tale. If Minister of the Interior Clifford Sifton could grant blocks of claims to men who'd never even seen the Yukon, Adney wrote, "Surely it would be a graceful act for him to yet do something for this man, who scorns to be a beggar." Anxious to promote the notion that a Canadian was at least partly responsible for the great rush to riches in its own territory, the federal government awarded Henderson $200 a month for life. Henderson was ready to set out on yet another prospecting expedition when cancer claimed him in 1933.

The men whose discovery started it all worked steadily on their claims until the summer of 1899. Exhausted millionaires George Carmack, Skookum Jim and Tagish Charlie packed up for a vacation to Seattle. The trip to the outside was a maddening experience for Carmack, who watched "civilization" undo his partners and his wife.

After losing control and hacking up a hotel's wooden banisters (so she could find her way to her room), Kate found herself in the Seattle city jail. Before the night was through, Skookum Jim was also behind bars. "So much for the debasing tendencies of great wealth and the firewater of the white man," the newspaper moralized. Carmack realized this was no way for a man to make his mark in polite society and become an influential member of the business community.

"I am disgusted with the whole outfit," Carmack wrote to his sister in California. "My, but the papers had a nice writeup. If Kate's trunks were here, I would ship her back to Dyea mighty quick."

Carmack bought out his partners and parted from Kate. He became a married man again in 1900—although he insisted that he'd never actually married Kate in the first place. Marguerite Laimme was his new bride. She was a seasoned gold-camp habitué of the South African and Australian gold rushes, who had become the prosperous owner of one of the infamous Dawson City "cigar stores," which were establishments that sold much more than smokes. Kate filed for divorce from Carmack and then sued for maintenance. The suit came to nothing. She returned to her Tagish family in the North. For the next 19 years, she lived in a ramshackle cabin at Carcross until influenza claimed her in 1920.

George Carmack's gold fever lingered on. The moustachioed prospector continued to mine in the Pacific Northwest, with more personal pleasure than financial success. In May 1922, as the featured speaker at Vancouver's Cabin of the Yukon Order of Pioneers, Carmack read the first chapter of his book in progress, the true story of the discovery on Rabbit Creek. After a round of applause, a resolution was passed naming Carmack the man who started the Klondike Gold Rush. The next morning, he complained of a cold. A week later, with Marguerite by his hospital bed, he died of pneumonia.

Skookum Jim lived out his years in the North with Tagish Charlie. In 1904, Skookum Jim sold his claims to a large mining company and trekked through the Yukon as a solitary prospector. With the help of church officials, he ensured the future needs of his wife and daughter and created a Whitehorse legacy for First Nations people known today as the Skookum Jim Friendship Centre. He died in 1916 at age 60. Tagish Charlie built a hotel at Carcross. Stumbling across a railway bridge after a Christmas celebration, Tagish Charlie, 42, fell to the water below and drowned.

Captain William Moore finally had his day in court, and when that day was done, judges awarded the visionary 25 percent of the value of Skagway's townsite. Moore was further enriched each time a boat tied up at the wharves that he had conceived a decade before the rush. Moore and his wife lived in northern splendour in a magnificent home that later became Skagway's world-famous Pullen House hotel.

The Rush is History
One year into the gold rush, the former tent town of Dawson City became a thriving city of 7,000 people, bigger than either Vancouver or Victoria. Business speculators were paying $20,000 for a prime corner building lot, a piece of ground that barely two years earlier had been worthless swamp. At three mills, saws whined constantly day and

night, and they still could not satisfy builders' insatiable appetite for lumber. Grand Forks grew large enough to merit its own police detachment, housed in a log building across from prostitute row.

Frenetic prospecting activity all but obliterated Bonanza and Eldorado Creeks, denuding hillsides and reshaping the topography, littering it with piles of rock. Many claim owners rarely visited the site, content to let gangs of workers toil away in the darkness and the dirt, surrounded by serpentine webs of elevated sluice channels.

Two years after the treasure ships carried the first wealth of the Klondike back to the world, paddlewheelers churning up the Yukon River to Dawson City arrived, carrying a cargo more precious than deckhands could unload down the gangways. The boats brought news of big gold strikes on the Bering Sea. The sands of the beaches of Nome, Alaska, were said to be laced with gold—lots of gold. In August 1899, within a single week, 8,000 people left Dawson City, never to return.

As quickly as it had begun, the Klondike Gold Rush was over.

Bibliography

Adney, Tappan. *The Klondike Stampede.* New York: Harper & Bros., 1900. Reprint, Vancouver: UBC Press, 1994.

Backhouse, Francis. *Women of the Klondike.* Vancouver: Whitecap Books, 1995.

Berton, Pierre. *Klondike: The Last Great Gold Rush 1896–1899.* Rev. ed. Toronto: McClelland & Stewart Ltd., 1975.

Bolotin, Norm. *Klondike Lost: A Decade Of Photographs by Kinsey & Kinsey.* Anchorage: Alaska Northwest Publishing Company, 1980.

Bolotin, Norm. *A Klondike Scrapbook.* San Francisco: Chronicle Books, 1987.

Bronson, William. *The Last Grand Adventure.* New York: McGraw-Hill Book Company, 1977.

Dobrowolsky, Helene. *Law Of The Yukon: A Pictorial History of the Mounted Police in the Yukon,* Whitehorse: Lost Moose Publishing, 1995.

Duncan, Jennifer. *Frontier Spirit: The Brave Women of the Klondike.* Toronto: Anchor Canada, 2004.

Gates, Michael. *Gold at Fortymile Creek.* Vancouver: UBC Press, 1994.

Horrall, S.W. *The Pictorial History of the Royal Canadian Mounted Police.* Toronto: McGraw-Hill Ryerson, 1973.

Johnson, James Albert. *George Carmack.* Vancouver: Whitecap Books, 2001.

Macdonald, Ian, and Betty O'Keefe. *The Klondike's "Dear Little Nugget."* Victoria: Horsdal & Schubart, 1996.

Bibliography

Mayer, Melanie J. *Klondike Women: True Tales of the 1897–1898 Gold Rush*. Athens, Ohio: Swallow Press, 1989.

Mayer, Melanie J. *Staking Her Claim: The Life of Belinda Mulrooney*. Athens, Ohio: Swallow Press, 2000.

Minter, Roy. *The White Pass: Gateway to the Klondike*. Toronto: McClelland & Stewart Ltd., 1987.

Morgan, Murray. *One Man's Gold Rush: A Klondike Album*. Seattle: University of Washington Press, 1973.

Neufeld, David. *Chilkoot Trail: Heritage Route to the Klondike*. Whitehorse: Lost Moose Publishing, 1996.

Ogilvie, William. *Early Days on the Yukon*. Whitehorse: Wolf Creek Books Inc., 2004.

Olive, W.H.T. *The Right Way On: Adventures in the Klondyke of 1898*. Langley: Timberholme Books, 1999.

Porsild, Charlene. *Gamblers and Dreamers: Women, Men, and Community in the Klondike*. Vancouver: UBC Press, 1998.

Steele, Samuel B. *Forty Years in Canada*. New York: Dodd, Mead & Co., 1915. Reprint, Toronto: Prospero Books, 2000.

Stewart, Robert. *Sam Steele: Lion of the Frontier*. 1979. Reprint, Regina: Centax Books, 1999.

Walden, Arthur T. *A Dog Puncher on the Yukon*. Boston: Houghton Mifflin Co., 1928. Reprint, Whitehorse: Wolf Creek Books, 2001.

Wallace, Jim. *Forty Mile to Bonanza: The North-West Mounted Police in the Klondike Gold Rush*. Calgary: Bunker To Bunker Publishing, 2000.

Wells, E. Hazard. *Magnificence and Misery: A Firsthand Account of the 1897 Gold Rush*. Edited by Randall M. Dodd. Garden City: Doubleday & Co., 1984.

Wright, Allen. *Prelude to Bonanza: The Discovery and Exploration of the Yukon*. Sidney, BC: Gray's Publishing Ltd., 1976.

Index

Index

Acknowledgements

The Klondike Gold Rush was a fitting climax to a tumultuous century, a period of significant social and technological advancements, wars and insurrections, and natural and man-made disasters. Yet, so significant was the Klondike Gold Rush that in the century that followed, no continental event—including Pancho Villa's Mexican revolution—rivalled its impact. With the exception of the American Civil War, the Klondike Gold Rush was *the* big story.

Over a century after it began, the rush continues to be among the most voluminously (and frequently inaccurately) reported events in modern history. Just as the sourdough separated "colour" from gravel at the bottom of a pan, the Klondike Gold Rush historian must separate fact from fiction in accounts written at the time and ever since. Like prospecting, it offers both challenge and satisfaction.

Like many other books on the Klondike rush, *Gold Fever* would not have been possible without Pierre Berton's *Klondike: The Last Great Gold Rush*. New discoveries made after its initial 1958 publication prompted Berton to revise his classic five years later. Since then, diligent research and lucky finds (such as Hazard Wells' diaries, Anna DeGraf's memoirs and George Carmack's papers and documents) continue to reveal new factual "nuggets," a phenomenon that Berton himself cheerfully acknowledged in the prefaces he provided for many later additions to the growing Klondike library.

The author wishes to thank Heritage House for the opportunity to expand and revise the original *Gold Fever*. Few authors have a second chance to improve upon a published book. The bibliography includes works that proved most helpful in preparing both the first and second editions.

About the Author

Author and freelance journalist Rich Mole has been a broadcaster, communications consultant and the president of a successful Vancouver Island advertising agency.

Rich is the author of numerous Klondike books, including *Murder and Mystery in the Yukon* and *Rebel Women of the Gold Rush*. Other non-fiction titles include *Christmas in British Columbia*, *Christmas in the Prairies*, and the hockey histories *Great Stanley Cup Victories* and *Against All Odds*, the story of the Edmonton Oilers.

Rich now lives in Calgary, where he is currently at work on a second novel. He can be reached at ramole@telus.net.

More Great Books in the Amazing Stories Series

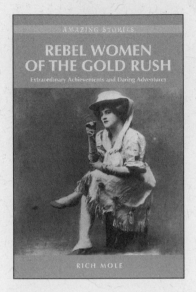

Rebel Women of the Gold Rush

Extraordinary Achievements and Daring Adventures

Rich Mole

(ISBN 978-1-894974-76-9)

Many of the women who arrived in the Klondike longed for the thrill of adventure and an end to the tedium of everyday life, while others sought riches. At a time when women were expected to conform to society's strict rules, the rebel women of the Klondike broke them with gusto, turning dreams into realities. They became millionaires, entrepreneurs, prostitutes, widows, wives, rebels and even murderers, scandalizing society in the process. Often on a trail of heartbreak and false hopes, they came to love the vast, untamed land that played a starring role in their own inspiring stories.

Visit www.heritagehouse.ca to see the entire list of books in this series.